W9-AXR-208

Selected as one of *The New York Times*'s
100 Notable Books of 2020 • A *New York Times Book
Review* Editors' Choice • Selected as one of *Vanity Fair*'s
Best Books of 2020 • A *Publishers Weekly* Best Nonfiction
Book of 2020 • A *Town & Country* Best Book of 2020

"Blume [is] a tireless researcher and beautiful writer, who moves through her narrative with seeming effortlessness—a trick that belies the skill and hard labor required to produce such prose. . . . *Fallout* is a warning without being a polemic . . . a book of serious intent that is nonetheless pleasant to read. There are knowable reasons for this, including Blume's flawless paragraphs; her clear narrative structure; her compelling stories, subplots and insights."

—William Langewiesche, *The New York Times*

"Blume's magisterial account of how John Hersey broke the story in the *New Yorker* is also a warning about the ever-present dangers of nuclear war."
—*The New York Times Book Review* (Editor's Choice)

"Gripping . . . Blume's meticulously researched tale of the lengths to which a government will go to keep the truth from reaching its citizens might be exactly what everyone should be reading at this deeply worrisome juncture."

—*The Washington Post*

"Totally riveting . . . It turns out that a lot of the issues that journalists are struggling with now, in terms of slowing down, telling a big story, telling a story of tragedy that resonates with people, was just as hard then as it is now. . . . I really loved this book."
—Kyle Pope, editor and publisher, *Columbia Journalism Review*

"[A] cliff-hanging saga of an intrepid young newsman outplaying his own government to get the facts."

—*The Wall Street Journal*

"Journalism at its finest . . . Blume's tight, fast-moving book, pegged to the 75th anniversary of the bombing, tells Hersey's story as he raced to gather sources, wrote in absolute secrecy, and then published a deeply empathetic, almost unbelievably distressing article."

—*Bloomberg*

"*Fallout* gives powerful insights into the way that a government can weave a story to justify the actions it takes, and also into the fearless reporting about what really happened in Hiroshima. Blume's tireless reporting gives important context to an understudied slice of U.S. history."

—*The Christian Science Monitor*

"As riveting as it is disturbing."

—*The Sydney Morning Herald*

"[A] brilliant book . . . [*Fallout*] powerfully shows how one courageous American reporter unraveled one of the deadliest cover-ups of the 20th century."

—Sara Z. Kutchesfahani, *The Bulletin of the Atomic Scientists*

"Meticulously researched and elegantly written . . . [*Fallout* is] an important reminder that the biggest stories may be hiding in plain sight."

—*The Nation*

"A book that deserves a wide readership."

—*The Oregonian*

"A brilliantly conceived and impeccably researched book, *Fallout* should sit next to [Hersey's *Hiroshima*] on the shelf as a testament of the courage of a free press to report the truth no matter who attempts to silence their mission."

—*New York Journal of Books*

"Blume skillfully reconstructs the hidden history of one of the greatest cover-ups in modern history . . . [a] thrilling account."

—*Town & Country*

"This enthralling, fine-grained chronicle reveals what it takes to cut through 'dangerously anesthetizing' statistics and speak truth to power."

—*Publishers Weekly* (Starred Review)

"A work of historical excavation . . . [Blume's] narrative never flags in its drama."

—*Kirkus Reviews* (Starred Review)

"[*Fallout* is] engagingly told and painstakingly researched, with an unerring eye for the vivid detail."

—*Booklist*

"Blume's work, like Hersey's, is a testament to the power of fine journalism. She brilliantly recreates his fragile position as the ultimate whistleblower, as well as his earth-shaking reporting."

—Bookreporter.com

"*Fallout* is gripping history. A big, important story; deeply researched and well told."

—Dan Rather

"In *Fallout*, Lesley Blume brilliantly tells the story of how John Hersey made his epic book *Hiroshima*, which had a profound effect on the way people came to regard atomic warfare. But the memory of his book has grown dim, and *Fallout* serves as an essential reminder of the lessons we once learned from Hersey's reporting."

—William J. Perry, 19th U.S. Secretary of Defense

"At a time when our world-destroying arsenal of nuclear weapons seems to have been all but forgotten, Lesley Blume's eloquent rediscovery of the story behind John Hersey's startling 1946 narrative 'Hiroshima' reminds us again of the vast human disaster even a small, primitive atomic bomb can visit upon the world."

—Richard Rhodes, Pulitzer Prize–winning author of
The Making of the Atomic Bomb

"In documenting how John Hersey pulled off one of the greatest journalistic feats in history, Blume has herself pulled off a great feat. *Fallout* is a fast-paced, deeply reported revelation."

—Gay Talese

"A searing testament to the power of journalism, truth-telling, and a story to help us remember our shared humanity . . . an urgent read."

—Sarah Sentilles, author of *Draw Your Weapons*

FALLOUT

The Hiroshima Cover-up
and the Reporter
Who Revealed It to the World

LESLEY M.M. BLUME

Simon & Schuster Paperbacks
New York   London   Toronto   Sydney   New Delhi

Simon & Schuster Paperbacks
An Imprint of Simon & Schuster, Inc.
1230 Avenue of the Americas
New York, NY 10020

Copyright © 2020 by Lesley M.M. Blume

All rights reserved, including the right to reproduce this book or portions thereof
in any form whatsoever. For information address Simon & Schuster Subsidiary
Rights Department, 1230 Avenue of the Americas, New York, NY 10020.

First Simon & Schuster trade paperback edition August 2021

SIMON & SCHUSTER PAPERBACKS and colophon are registered
trademarks of Simon & Schuster, Inc.

For information about special discounts for bulk purchases, please contact
Simon & Schuster Special Sales at 1-866-506-1949 or business@simonandschuster.com.

The Simon & Schuster Speakers Bureau can bring authors to your live event.
For more information or to book an event contact the Simon & Schuster Speakers Bureau
at 1-866-248-3049 or visit our website at www.simonspeakers.com.

Interior design by Lewelin Polanco

Manufactured in the United States of America

1 3 5 7 9 10 8 6 4 2

Library of Congress Cataloging-in-Publication Data

Names: Blume, Lesley M. M., author.
Title: Fallout / Lesley M. M. Blume.
Description: First Simon & Schuster hardcover edition. |
New York : Simon & Schuster, 2020. | Includes bibliographical references and index. |
Summary: "New York Times bestselling author Lesley M.M. Blume reveals how a courageous
reporter uncovered one of greatest and deadliest cover-ups of the 20th century—the true effects
of the atom bomb—potentially saving millions of lives"—Provided by publisher.
Identifiers: LCCN 2020000055 (print) | LCCN 2020000056 (ebook) | ISBN 9781982128517
(hardcover) | ISBN 9781982128531 (paperback) | ISBN 9781982128555 (ebook)
Subjects: LCSH: Hiroshima-shi (Japan)—History—Bombardment, 1945—Press coverage—
United States. | World War, 1939-1945—Japan—Hiroshima-shi—Press coverage—United
States. | Atomic bomb victims—Japan—Hiroshima-shi—Press coverage—United States. |
Hersey, John, 1914–1993. | Hersey, John, 1914–1993. Hiroshima. | Journalists—United States—
Biography. | World War, 1939-1945—Press coverage—Japan. | Atomic bomb—United States—
Public opinion. | Atomic bomb—Government policy—United States—History—20th century.
Classification: LCC D767.25.H6 B58 2020 (print) | LCC D767.25.H6 (ebook) |
DDC 940.54/2521954—dc23
LC record available at https://lccn.loc.gov/2020000055
LC ebook record available at https://lccn.loc.gov/2020000056

ISBN 978-1-9821-2851-7
ISBN 978-1-9821-2853-1 (pbk)
ISBN 978-1-9821-2855-5 (ebook)

For Koko Tanimoto Kondo

What has kept the world safe from the bomb since 1945 has been the memory of what happened at Hiroshima.

—JOHN HERSEY

Contents

Introduction

John Hersey later claimed that he had not intended to write an exposé. Yet, in the summer of 1946, he revealed one of the deadliest and most consequential government cover-ups of modern times. The *New Yorker* magazine devoted its entire August 31, 1946, issue to Hersey's "Hiroshima," in which he reported to Americans and the world the full, ghastly realities of atomic warfare in that city, featuring testimonies from six of the only humans in history to survive nuclear attack.

The U.S. government had dropped a nearly 10,000-pound uranium bomb—which had been dubbed "Little Boy" and scribbled with profane messages to the Japanese emperor—on Hiroshima a year earlier, at 8:15 a.m. on August 6, 1945. None of the bomb's creators even knew for certain if the then experimental weapon would work: Little Boy was the first nuclear weapon to be used in warfare, and Hiroshima's citizens were chosen as its unfortunate guinea pigs. When Little Boy exploded above the city, tens of thousands of people were burned to death, crushed or buried alive by collapsing buildings, or bludgeoned by flying debris.

Those directly under the bomb's hypocenter were incinerated, instantaneously erased from existence. Many blast survivors—supposedly the lucky ones—suffered from agonizing radiation poisoning and died by the hundreds in the months that followed.

The city of Hiroshima initially estimated that more than 42,000 civilians had died from the bombing. Within a year, that estimate would rise to 100,000. It has since been calculated that as many as 280,000 people may have died by the end of 1945 from effects of the bomb, although the exact number will never be known. In the decades since, human remains have been regularly uncovered in the city's ground, and are still uncovered today. "You dig two feet and there are bones," says Hiroshima Prefecture governor Hidehiko Yuzaki. "We're living on that. Not only near the epicenter [of the blast], but across the city."

It was a massacre of biblical proportions. Even today—seventy-five years after the bombing—the name Hiroshima conjures up images of fiery nuclear holocaust and sends chills down spines around the world.

However, until Hersey's story appeared in the *New Yorker*, the U.S. government had astonishingly managed to hide the magnitude of what happened in Hiroshima immediately after the bombing, and successfully covered up the bomb's long-term deadly radiological effects. U.S. officials in Washington, D.C., and occupation officials in Japan suppressed, contained, and spun reports from the ground in Hiroshima and Nagasaki—which had been attacked by the United States with the plutonium bomb "Fat Man" on August 9, 1945—until the story all but disappeared from the headlines and the public's consciousness.

At first, the government appeared to be forthright about its new weapon. When U.S. president Harry S. Truman announced to the world that an atomic bomb had just been dropped on Hiroshima, he pledged that if the Japanese did not surrender, they could "expect a rain of ruin from the air, the like of which has never been seen on this earth." Little Boy had packed an explosive payload equivalent to more than 20,000 tons of TNT, the president revealed, and was by far the largest bomb ever used in the history of warfare. Reporters and editors given text of this presidential announcement in advance received the news with disbelief. Young Walter Cronkite—then a United Press war reporter based in Europe—upon receiving a bulletin from Paris about the bomb, thought that "clearly . . . those French operators [had] made a mistake," he recalled later. "So I changed the figure to 20 tons." Soon, as updates to the story came in, "my mistake became abundantly clear."

Also, it seemed at first that the press was adequately reporting on the fates of Hiroshima and Nagasaki. As the implications of the world's entrance into the atomic age began to sink in, it became apparent to editors and reporters everywhere that the atomic bomb was not just one of the biggest stories of the war but among the biggest news stories in history. After millennia of contriving increasingly horrible and efficient killing machines, humans had finally invented the means with which to extinguish their entire civilization. Humankind was "stealing God's stuff," as E. B. White wrote in the *New Yorker*.

Yet it would take many months—and the bravery of one young American reporter and his editors—before the world learned what had actually transpired beneath those roiling mushroom clouds. "What happened at Hiroshima is not yet known," reported the *New York Times* on August 7, 1945. "An impenetrable cloud of dust and smoke masked the target area from reconnaissance planes." In many respects, the impenetrable cloud didn't truly lift until Hersey got into Hiroshima in May 1946 and, weeks later, managed to publish an account of his findings there. Even though the *New York Times* was the only publication that had a reporter accompany the Nagasaki atomic bombing run and had maintained a bureau in Tokyo since the Japanese surrender, *Times* reporter (and later managing editor) Arthur Gelb stated that "most of us were unaware, at first, of the extent of the devastation caused by the bombs. John Hersey's excruciatingly detailed account . . . finally brought home to Americans the magnitude of the event."

Media coverage of the bombings had been initially widespread and intensive, but details of the aftermath were actually scarce from the beginning, thanks to U.S. government and military efforts to control information about their handiwork in Hiroshima and Nagasaki. The United States—which had just won a painfully earned moral and military victory over the Axis powers—was not eager to "get the reputation for outdoing Hitler in atrocities," as the country's secretary of war put it. Right away, officials in Washington, D.C., and newly arrived occupation forces in Japan went into overdrive to contain the story of the human cost of their new weapon. The Japanese media was forbidden by occupation authorities to write or air stories about Hiroshima or Nagasaki, lest they "disturb public tranquility." As foreign reporters began to get into the country, Hiroshima and Nagasaki were immediately put off-limits to them. The few journalists attempting to report on the atomic cities in the weeks immediately following the bombings were threatened with expulsion from Japan, harassed by U.S. officials, and accused of spreading Japanese propaganda, dispensed by a defeated enemy attempting to cultivate international sympathy after years of aggression and their own outsized atrocities.

On the home front, U.S. government officials corralled the population into thinking of the atom bomb as a conventional superbomb, painting it in terms of TNT and denying its radioactive aftermath. "It was just the same as getting a bigger gun than the other fellow had to win a war and that's what it was used for," said President Truman. "Nothing else but an artillery weapon." When it was eventually conceded that bomb-induced radiation poisoning was real, its horrors were downplayed. (It could even be a "very pleasant way to die," stated Lieutenant General Leslie R. Groves, head of the Manhattan Project, which had created the bombs in just three years.)

The American public was allowed to see images of the mushroom clouds and hear triumphant eyewitness descriptions from the American bombers themselves, but reports containing testimonies from below the clouds were virtually nonexistent. Images of Hiroshima's and Nagasaki's devastated landscapes were also released to newspapers and magazines by U.S. forces. However, while sobering, the post-atomic landscape photographs failed to register deeply enough with readers who had been inundated with images of decimated cities—London, Warsaw, Manila, Dresden, Chungking, among scores of others—on a daily basis for more than half a decade. Hersey himself acknowledged that post-bomb landscape photos could only get a limited emotional response; ruins, he thought, could be "spectacular; but . . . impersonal, as rubble so often is." What the American public did not see: photos of the Hiroshima and Nagasaki hospitals ringed by the corpses of blast survivors who had staggered there seeking medical help and died in agony on the front steps. (Most of the doctors and nurses had been killed or wounded anyway.) Nor did they see images of the crematoriums burning the remains of thousands of anonymous victims, or pictures of scorched women and children, their hair falling out in fistfuls.

The published images of Hiroshima's demolished landscape gravely undersold the reality of atomic aftermath. Usually a picture is worth a thousand words, but in this case it would take Hersey's 30,000 words to reveal and drive home the truth about America's new mega-weapon. The Japanese, of course, didn't need Hersey to educate them about the effects of Little Boy and Fat Man, but American readers were shocked when they were, at last, properly introduced to the nuclear bombs that had been detonated in their name.

* * *

Fallout is the backstory of how John Hersey got the full story about atomic aftermath when no other journalist could, and how "Hiroshima" became—and remains—one of the most important works of journalism ever created. Over the past seven decades, Hersey's "Hiroshima" has not, of course, prevented dangerous nuclear arms races; nor have its revelations solved the problems of the atomic age, just as the *Washington Post*'s Watergate reporting did not solve the problem of government corruption.

But as the document of record—read over the years by millions around the world—graphically showing what nuclear warfare truly looks like, and what atomic bombs do to humans, "Hiroshima" has played a major role in preventing nuclear war since the end of World War II. In 1946, Hersey's story was the first truly effective, internationally heeded warning about the existential threat that nuclear arms posed to civilization. It has since helped motivate generations of activists and leaders to work to prevent nuclear war, which would likely end the brief human experiment on earth. We know what atomic apocalypse would look like because John Hersey showed us. Since the release of "Hiroshima," no leader or party could threaten nuclear action without an absolute knowledge of the horrific results of such an attack. That is, unless that act was one conducted amidst willful ignorance—or nihilistic brutality.

Casualty statistics can be numbing. While the initial lack of comprehension in the United States over Hiroshima's fate was largely due to the government's active suppression of information from the ground there, it did not help that much of the population was suffering from atrocity exhaustion by the end of the war. By 1946, Americans had been witnesses—along with the rest of the world—to carnage on an unprecedented scale. World War II remains the deadliest conflict in human history. The National WWII Museum estimates that, worldwide, 15 million combatants died, along with some 45 million civilians—although there may have been as many as 50 million civilian casualties among the Chinese alone. Russia puts its losses at 26.6 million dead; the United States lost more than 407,000 military servicemen and women. Every day during the war, gruesome death toll statistics were announced in

American publications from fronts around the globe. The more zeros attached to a statistic, the more unfathomable it was. Somewhere along the way, the numbers seemed to stop representing the bodies of actual people; the human element became divorced from the tallies.

In "Hiroshima," Hersey informed his readers that 100,000 had died thus far in that atomic city as the result of the bombing. Yet had he presented this number and his other findings in a straightforward news story, "Hiroshima" likely would not have had such a visceral and enduring impact. As one of Hersey's journalist contemporaries, Lewis Gannett of the *New York Herald Tribune*, put it, "When headlines say a hundred thousand people are killed, whether in battle, by earthquake, flood, or atom bomb, the human mind refuses to react to mathematics." In the immediate aftermath of the bombings, Americans were given varying estimates of Hiroshima and Nagasaki casualties—all of them grotesquely high, especially when one remembered that a single bomb was responsible for all of that death—but to no avail.

"You swallowed statistics, gasped in awe," Gannett wrote, "and, turning away to discuss the price of lamb chops, forgot. But if you read what Mr. Hersey writes, you won't forget."

For Hersey, driving home the gruesome reality behind those impersonal numbers was essential. Since 1939 he had covered various battlefronts and seen the savagery of which humans of all nationalities were capable once they stopped seeing their enemies and captives as fellow human beings. The best chance that mankind had for survival—especially now that warfare had gone nuclear, Hersey felt—was if people could be made to see the humanity in each other again.

This was a tall order. To create a work that would help restore a shared sense of humanity, Hersey would not only have to get behind those dangerously anesthetizing stats but also tackle the virulent, reductive racism that had given rise to wartime genocides and atrocities around the globe. Humanizing the Japanese for an American audience would be especially controversial and difficult. Hatred and suspicion toward the Japanese ran deep in this country after Pearl Harbor. "American pride [had] dissolved overnight into American rage and hysteria," Hersey recalled later. Approximately 117,000 people of Japanese descent had been detained in internment camps in the United States during the war. Hollywood had long been hard at work churning out propaganda and feature films warning of the subhuman yellow peril from the east. News about cruelties inflicted on American prisoners of war during the 1942 Bataan Death March, Japanese atrocities committed against civilians in China, and the savage battles over atolls in the Pacific had horrified Americans and reinforced the idea that all Japanese were bestial and fanatical.

In his speech announcing the Hiroshima bombing, President Truman had spoken for many Americans when he stated that, with the atomic attack, the Japanese "have been repaid many fold" for their own attack on Pearl Harbor four years earlier. The citizens of Hiroshima and Nagasaki had gotten what they deserved; it was as simple as that. One poll conducted in mid-August revealed that 85 percent of those surveyed endorsed the bombs' use, and in a different poll around that time 23 percent of those surveyed regretted that the United States didn't get a chance to use "many more of the bombs before Japan had a chance to surrender." Hersey had seen firsthand in Asia and the Pacific evidence of Japanese barbarity and tenacity in battle. Still, he was determined to make sure Americans could see themselves in the citizens of Hiroshima.

"If our concept of . . . civilization was to mean anything," he stated, "we had to acknowledge the humanity of even our misled and murderous enemies."

When he got into Japan, and then into Hiroshima—no small feat in an occupied country closely controlled by General Douglas MacArthur and his forces—Hersey managed to interview dozens of blast survivors. Among them: a struggling Japanese widow with three young children;

a young Japanese female clerk; two Japanese medics; a young German priest; and a Japanese pastor. In his story for the *New Yorker*, Hersey recounted—in minute, painful detail—the day of the bombing from each of these six survivors' point of view.

"They still wondered why they lived when so many others died," Hersey wrote. That day and since, each had seen "more death than he ever thought he would see."

Through their eyes, Hersey also made Americans see more death than they ever thought they would see—and a new, uniquely awful version of death at that. As people read "Hiroshima," they visualized New York or Detroit or Seattle in Hiroshima's stead, and imagined their own families and friends and children enduring the same hell on earth. Just as Hersey had managed to access Hiroshima itself against the odds, he had successfully breached the fatigue—and tribal barriers—and broken them down. Almost miraculously, he had managed to trigger empathy.

The simplicity of his approach—premised on portraying six relatable people whose lives were violently upended at the same moment—mirrored the basic power of the tiny, mighty atom itself.

* * *

The U.S. government's attempt to suppress information about Hiroshima had been almost ridiculous, Hersey felt; equally absurd was the government's bid to retain its initial nuclear monopoly. Sooner or later (and likely sooner, he thought) other countries were bound to figure out the physics, and it was only a matter of time before the truth about Hiroshima and Nagasaki got out. Yet, before he had personally gotten into Japan—ten months after the bombings—the American media had already essentially given up on trying to break the story of Hiroshima in a significant way, essentially giving Hersey an unlikely monopoly on the story.

Hersey's article had been released into a frenetic news landscape, with hundreds of stories and international developments vying for reporters' and the public's attention. The American press corps was in relentless pursuit of the next scoop, obsessed with getting the edge on the next big story. Dozens of foreign correspondents had been dispatched by their news organizations to Tokyo since the Japanese surrender a year earlier. Occupation authorities had indeed largely managed to squelch the few bold early attempts to cover Hiroshima and Nagasaki, and they closely monitored and controlled Japan-based reporters after that point. Yet, as time went on, many of Hersey's reporter colleagues had started to lose interest in reporting on Hiroshima's fate anyway; it started to seem like yesterday's news, and they directed their attention to other stories. Back home, their editors were quietly asked to submit press reports about nuclear matters to the War Department; failure to do so could compromise national security, they were advised. They largely complied.

The *New Yorker*'s founder and editor, Harold Ross, had directed his wartime writers to find consequential, hiding-in-plain-sight stories ignored by other reporters. Hersey took note, and when the *New Yorker* released "Hiroshima," the story not only had the feel of an exposé, but it appeared to be the scoop of the century. (The story had certainly been treated that way in-house at the magazine: Ross and his managing editor, William Shawn, kept the "Hiroshima" project strictly under wraps—going to almost absurdly dramatic lengths to keep it secret even from the magazine's own staffers—until just before the article's release.) When Hersey's story came out, the media reaction was frenzied: "Hiroshima" made front-page news around the world and was covered on more than five hundred radio stations in the United States alone—even though Hersey's feat revealed that every other press outlet had actually missed the huge story that they had *seemed* to cover so diligently.

The public relations fallout created by "Hiroshima" also embarrassed the U.S. government, which scrambled to contain the damage. But once

"Hiroshima" ran in the *New Yorker*, the genie could not be put back into the bottle. Now that the cover-up was blown, the reality of nuclear aftermath was a matter of permanent, policy-influencing international record. Hersey had made it impossible for Americans to avert their eyes and, as physicist Albert Einstein put it, "escape into easy comforts" again.

That said, the Manhattan Project's General Leslie Groves—who had played a central early role in distorting and hiding information about Hiroshima and the weapon he'd helped create—did play a surprising role in bringing "Hiroshima" to the masses. And the U.S. government and military would find their own unlikely and cynical utility in the article once it had been published. While Hersey's article had indeed embarrassed the United States, some government figures realized that it wasn't entirely a bad thing that "Hiroshima" had showcased, to great effect, the devastating power of the United States' new weapon—a most unwelcome reminder to America's rivals, who were still years away from developing their own nuclear weapons. (To that end, the Soviets deeply resented "Hiroshima" and its author; their hostility became increasingly vehement over time. Actions were taken in Russia to debunk Hersey's revelations, smear Hersey himself, and downplay the might of America's new bombs.) In retrospect, the "Hiroshima" story reveals much about the U.S. government's internal conflict over how much to showcase about the atomic bomb and how much to hide about it at all costs.

Whatever import "Hiroshima" took on in various realms, Hersey and his editors at the *New Yorker* always saw the article as a document of conscience. Also released almost immediately in book form around the world and in many languages, "Hiroshima"—with its continued ability to engulf readers emotionally—has sold millions of copies and long acted as a pillar of deterrence. Years later, Hersey would comment on the role that such eyewitness testimonies had played in keeping subsequent generations of leaders from incinerating the planet. It "has not been deterrence, in the sense of fear of specific weapons," he said, "so much as it's been memory. The memory of what happened at Hiroshima."

* * *

Most journalistic works have short shelf lives. Yet "Hiroshima" is dated in only one respect: the story's hell-wreaking main character, Little Boy, was already considered primitive by the time Hersey wrote his 1946 story just months after the bomb's detonation. The United States had already begun developing the hydrogen bomb, which would prove many times more powerful than the atomic bombs dropped on Japan. Today's nuclear arsenals include hundreds of bombs vastly more powerful than Little Boy or Fat Man. (The most powerful nuclear device—called the Tsar Bomba, detonated by the Soviets in 1961—was reportedly 1,570 times more powerful than the yield of the bombs dropped on Hiroshima and Nagasaki combined, and ten times more powerful than all of the conventional weapons exploded during World War II.) It is estimated that the world's current combined inventory of nuclear arms includes approximately 13,000 warheads. Should war break out today, the prognosis for civilization's survival is grim; as Einstein said after the Japan bombings: "I do not know how the Third World War will be fought, but I can tell you what they will use in the Fourth—rocks."

Rampaging climate change and pandemic have been cited as *the* immediate existential threats to human survival; yet nuclear weapons continue to pose another urgent existential threat—and according to nuclear experts, that threat has never been more imminent. The onset of climate change has been violent and frightening, yet its effects are relatively gradual. Nuclear war, on the other hand, could spell instantaneous global destruction, with little or no advance warning. Hersey had, in the 1980s, worried about "slippage"—a hair-trigger mistake or misinterpretation between two nuclear powers that could lead to an immediate, irreversible nuclear confrontation. If such "slippage" occurred now, leaders could, in a matter of minutes, incite events that would wipe out all life on earth.

Introduction

In recent years, long-standing barriers to such nuclear conflagrations have been weakened. Leaders of nuclear-armed nations accelerated production of and modernized their nuclear arsenals. International treaties restricting such escalation were abandoned as channels of communication among leaders of nuclear powers deteriorated. North Korea provocatively tested atomic missiles and weapons while the Trump administration vacillated between provoking North Korean leader Kim Jong-un, ineptly cultivating him, and looking the other way. For the second year in a row, the *Bulletin of the Atomic Scientists*, a nuclear watchdog group, has set its Doomsday Clock—which gauges the world's proximity to the possibility of nuclear war—to "100 seconds to midnight," with midnight meaning nuclear apocalypse. Before 2020, the clock had never been set that close to doomsday—not even in 1953, "the most dangerous year of the Cold War," says Dr. William J. Perry, former U.S. secretary of defense and chair of the *Bulletin*'s board of sponsors. "Most people don't understand that the dangers we are facing today are equivalent to those dark days."

Experts maintain that climate change is contributing to this dangerous nuclear landscape: civil wars sparked in part by environmental upheaval are a factor in forcing refugee movements in record numbers, exacerbating tensions among nations and prompting populations around the globe to elect militaristic leaders. To make matters even worse, the sort of virulent nationalism and racism that helped set the stage for World War II—and which Hersey had worked so hard to break down with "Hiroshima"—has been surging again throughout the world, with much of it fueled by social media. Americans are proving far from immune to this trend of dehumanization; for example, many have indicated that they would now be willing to inflict extreme mass casualties on civilians of an enemy state via preemptive nuclear attack. A recent survey of 3,000 Americans revealed that a third of those surveyed supported such a strike, even if that meant a million North Korean civilians would die in the attack. "It's our best chance of eliminating North Koreans," stated one strike supporter. The purpose of the strike, according to another: "to end North Korea."

In 1946, Hersey wrote that his protagonists did not yet understand why they had survived the Hiroshima bombing while tens of thousands of others around them had perished. Part of the reason, Hersey felt, was to warn future generations about the cruel impact of a bomb that continues to kill long after it is detonated, and to help ensure that nuclear weapons are never used again. He hoped that his documentation of Hiroshima's fate would continue to serve as a deterrent. But if the lesson of Hiroshima was ignored or forgotten, he warned, continued human existence was indeed a "Big If."

Chapter One

The Picture Does Not Tell the Whole Story

LIMBO

New York City, May 8, 1945. Victory in Europe Day, or V-E Day. German forces in Europe had just surrendered unconditionally to the Allies. Hitler had killed himself a week earlier. After years of bloodshed and destruction, the war in Europe was over at last.

A quarter of a million of people crowded into Times Square. Over a thousand tons of paper—ripped newsprint, torn telephone book pages, anything shreddable—showered down from windows in the surrounding buildings into the streets below. On Wall Street, a blizzard of ticker tape swirled in the air. Boats on the Hudson and East Rivers blew their horns, which mingled with the cheering on land to produce a joyous, deafening cacophony.

John Hersey had more than one reason to celebrate that day. Not only was he likely exulting along with other New Yorkers about the end of hostilities in Europe—which he had covered on various fronts as a war correspondent—but he also received some very good personal news. He and his friend Richard Lauterbach, a correspondent for *Time* and *Life* magazines, were playing tennis at Rip's Tennis Courts in midtown Manhattan, near the East River, tucked away from the Times Square revelries. One of the court staffers came out of the club shed onto the court and hollered at Hersey.

"I just heard on the radio that you won the Pulitzer Prize," he said.

Hersey didn't believe it. After a beat, he turned to his friend on the court.

"Lauterbach, you bastard, you're trying to pull a fast one on me," he told him. "I know it!"

Lauterbach apparently didn't try to dissuade Hersey that he was being pranked. The men played out the rest of the set. When, later that day, Hersey returned home to his Park Avenue apartment, where he lived with his wife and their three young children, he discovered that he actually *had* won a Pulitzer Prize for his 1944 novel, *A Bell for Adano*.

Even before this accolade, Hersey—just thirty years old when he won the Prize—had already had an enviable career. A respected international correspondent for *Time* magazine throughout the war, he was also a war hero. The secretary of the navy had personally sent him a note of commendation after Hersey had helped evacuate wounded Marines while on assignment covering a battle between Japanese and Allied forces in the Solomon Islands. ("I should have sent it back," Hersey later said. "My alacrity in helping to get the wounded out was my way of taking the quickest possible exit from that hellhole.") Before *A Bell for Adano* was published in 1944, he had already authored two other well-received books: *Men on Bataan* (1942), a biography of General Douglas MacArthur and his forces, who had since been painfully fighting their way, island by island, up through the Pacific toward Japan; and *Into the Valley* (1943), which depicted the bloody dogfight he'd survived on Guadalcanal. Even before earning Hersey the Pulitzer, *A Bell of Adano*—which depicted a Sicily-based American major who tries to help locals find a replacement for the seven-hundred-year-old town bell that had been melted down for bullets by Fascists—had already been adapted into both a movie and a Broadway play.

Once the Pulitzer was bestowed upon him, Hersey's literary star rose even higher. Critics compared him to Hemingway. He and his wife, Frances Ann—a wealthy, educated Southern-born-and-raised beauty who had been presented at the Court of St. James's in London—were leading a glamorous life. The film version of *A Bell for Adano* was released just weeks after V-E Day, in June. There was an invitation to the White House; powerful gossip columnist Walter Winchell mentioned him in his column.

Despite the fanfare, however, Hersey maintained a relatively low profile and an attractive sense of humility. For years, friends and colleagues would cite that humbleness as one of his defining characteristics and puzzle over its origin. After all, he had been almost excessively celebrated throughout his life. Enrolled as a scholarship student at The Hotchkiss School—a posh Connecticut boarding school—he was, in his senior year, voted "most popular member of the class" and also the "most influential." When he moved on to Yale University, he was tapped for the exclusive Skull and Bones society, which boasted presidents, diplomats, and publishing moguls among its alums.

The humility may have come from his early background: Hersey had been born in China to American missionaries. While not religious himself, his reserve and pronounced moral compass were likely rooted in that upbringing, along with his staunch aversion to self-promotion. Amidst the acclaim of his early career, Hersey would find personal attention "hollow," one of his sons would recall later, and developed an early antipathy to "flogging his wares." As Hersey's career developed, he always preferred instead to "let his works speak for themselves," added one of his daughters. He lived in the spotlight and yet he seemed—to the public, anyway—something of a cipher. This suited him just fine.

Despite his celebrity that summer, Hersey was at a professional crossroads. He had recently returned to the United States from Moscow, where he had opened the *Time* bureau in 1944 after covering various theaters of war for that magazine since 1939. It had been a frustrating, complicated assignment. Hersey had been at loggerheads not only with his Soviet hosts but also with his boss, Time Inc. cofounder and editor Henry Luce. The Soviets had confined and monitored Hersey and the other Western correspondents based in Moscow; he and his fellow reporters had, Hersey remembered, spent most of their time drinking at the Metropol hotel while trying "to catch a glimpse of the war, which was several hundred miles away."

Luce, for his part, despised the Soviets—then wartime allies of the United States—and communism. In his opinion the twentieth century belonged rightfully to America, democracy, and free enterprise. He and his top editor in New York rarely printed anything that Hersey wrote from the Russian capital, and when they did, they rewrote and edited Hersey's stories so egregiously that Hersey grew angry and threatened to quit; at one point he reportedly told Luce to his face that "there was as much truthful reporting in *Pravda*"—then the mouthpiece of the Soviet government—as there was in *Time*. This relationship deterioration was a regrettable development for Luce, who—despite muzzling the Russia dispatches—had actually hoped to groom Hersey for a leadership role in Time Inc.'s expanding and influential magazine empire.

The *Time* boss had long been somewhat narcissistically fixated on Hersey. The two men shared bizarrely similar backgrounds: like Hersey, Luce had been born in China to American missionary parents (making them "mishkids," as Hersey put it); and like Hersey, he had been educated as a scholarship student at Hotchkiss and Yale. The one nominal digression in their educational résumés: Luce had undertaken postgraduate studies at Oxford University, and Hersey at Cambridge.

For Hersey, Luce had seemed, at first, a "walking wonder of possibilities," although he later downgraded the nature of the relationship to "quasi-parental." When he made it clear that he intended to quit, Luce panicked and tried to lure Hersey home to begin training him for *Time*'s managing editorship. The eleventh-hour seduction attempt failed. Hersey resigned on July 11, 1945, and returned to New York.

As the summer of 1945 stretched before him, Hersey was evaluating his options. He was now a freelancer instead of heir apparent to a publishing empire. Many of his journalist friends and colleagues remained overseas, covering the winding down of Hitler's defeated killing machine and the aftermath of the European conflict. The Pacific war continued to rage, and a feeling of queasy anxiety quickly settled back over New York City. Even during the V-E Day celebrations there, the shadow of still-undefeated Japan soured the festivities. Some revelers had tried to put a good face on the specter, carrying signs proclaiming:

"On to Tokyo!"

"On to Japan!"

"Two down, One to Go!"

The prognosis for beating Japan was at once encouraging and grim. That country's navy had been devastated; the Allies had gained territorial footholds from which they could conduct air raids over the Japanese mainland. Late that winter, a firebombing air raid over Tokyo had burned 16 square miles of the Japanese capital in a single night. Yet the Japanese showed no apparent sign of surrendering. Hersey, like many other Americans, feared that a Japanese land invasion would be necessary, with horrific casualties on both sides.

"I had been under fire in skirmishes against the Japanese, and had come to know how very tenacious and how very dedicated they were," he said.

The U.S. War Department had announced that it would begin diverting veterans of the European campaigns to the Pacific. Many of Hersey's fellow war correspondents now flocked to cover the Pacific campaigns as well, and embedded with Allied forces there. Among them was Bill Lawrence of the *New York Times*, who had been posted in Moscow with Hersey. Lawrence wrote to his editors and to Hersey about his different assignments, keeping them in the loop from afar. He and Hersey had been drinking buddies in Russia; Lawrence was a "bear of a man, lusty, the darling of the Katinkas" who had once passed out at a banquet in Leningrad and had to be dragged feetfirst out of the hall.

Lawrence's new assignment—covering the Allied invasion of Okinawa—was far more sobering. The fighting had been slow and excruciating, he reported back to New York; on the island he had witnessed U.S. aircraft spraying cave-filled hills with napalm and igniting the areas in what "the G.I.s called . . . 'Jap Barbecues.'" Otherwise, the fighting would have involved cave-to-cave, hand-to-hand combat. In Lawrence's opinion, the war with the Japanese was bound to last for years, and he saw no evidence that the will of Japanese soldiers was weakening. The U.S. military was preparing for an amphibious assault on Japan for the fall of 1945.

"Few of us in the Pacific knew . . . that our war was about to end," Lawrence later recalled. By mid-July, back in the United States, the first atomic bomb in history had been successfully—and secretly—detonated in the New Mexico desert; the bombs ultimately destined for Hiroshima and Nagasaki were being prepared.

A NEW AND MOST CRUEL BOMB

On August 6, 1945, Hersey was in Cold Spring Harbor, New York, when he heard President Truman announce on the radio that the United States had used an atomic bomb on Hiroshima. This new weapon, the president declared, drew its terrible power from the basic powers of the universe. "The force from which the sun draws its power has been loosed against those who brought war to the Far East," he said. If the Japanese did not submit unconditionally to the surrender terms already issued by Allied leaders the previous month at the Potsdam Conference, they could expect obliteration. More atomic bombs were in development, Truman advised, including even more powerful versions. The United States would continue to drop them, one after another, he said, until Japan capitulated.

Unlike Bill Lawrence, Hersey actually had heard about atomic bombs while still at *Time* magazine, so the news wasn't as bewildering to him as it was for almost everyone else. Most of the country and world had been kept in the dark about the $2 billion nuclear undertaking to create these nuclear weapons. Tens of thousands of people had worked on the Manhattan Project in covert locations across the United States without knowing exactly what they were constructing. American pilots had trained in Utah and the Pacific for a bombing mission whose details and goal were unknown to them: they had not "the slightest inkling of the nature of their job," recalled one observer at Tinian, the Pacific island base from which the Hiroshima bombardment team had taken off. "All of them had been asked to volunteer for an organization that was 'going to do something different.' That was all." President Truman hadn't even known about the project until the death of his predecessor, President Franklin D. Roosevelt, in April 1945, a mere three months before the first bomb was successfully tested in New Mexico.

Upon hearing the news about Hiroshima, Hersey was immediately overwhelmed by a sense of despair. It wasn't a feeling of guilt—or even, at first, compassion for Hiroshima's victims—but rather an overarching fear about the world's future. It was instantly clear to him that humanity had suddenly entered a terrifying new chapter. Yet he also felt relieved: the Hiroshima bomb—as horrible as it must have been, and as worrisome in its implications—would likely end the war at last.

His relief disintegrated three days later when the United States dropped a second atomic bomb on Japan, this time on the port city of

Nagasaki. Hersey was appalled. This second nuclear attack was an indefensible excess, in his opinion, a "totally criminal" action that resulted in tens of thousands of unnecessary deaths.

"We gave the Japanese a demonstration that was terrible," he later recalled, adding that he felt "sure that one bomb would have brought the Japanese surrender." The incendiary raids on cities in Japan—and Germany—had already seemed morally reprehensible to him, but the atomic bomb had just added "a terrifying factor of efficiency" to humanity's ability to inflict mass casualties in warfare.

Publications around the world began to print photographs of the ghoulish mushroom clouds that had appeared over Hiroshima and Nagasaki. A *New York Times* reporter who had accompanied the Nagasaki bombing run described the cloud emerging from that obliterated city as a "living totem pole, carved with many grotesque masks grimacing at the earth." From the vast mushroom cloud emerged a smaller mushroom cloud, "as though the decapitated monster was growing a new head." The bombing crew could still see the cloud from 200 miles away.

Now the world was waiting to hear and see what Hiroshima and Nagasaki looked like on the ground. "An impenetrable cloud of dust and smoke masked the target area from reconnaissance planes," the *New York Times* reported on August 7, and therefore "what happened at Hiroshima is not yet known. The War Department said it 'as yet was unable to make an accurate report.'"

Allied correspondents and editors awaited the initial reports on the fate of those in Hiroshima and Nagasaki. Those based in the Pacific monitored Japanese press and radio stations for any dispatches describing the fate of the atomic cities. But the Japanese media had been instructed by Japanese intelligence to downplay the attacks. ("Hiroshima was attacked by incendiary bombs," read one article in *Asahi Shimbun*, one of Japan's largest newspapers. "It seems that some damage was caused to the city and its vicinity.") The initial response in the publications was so subdued that U.S. officials worried that the Japanese had not yet fully comprehended their situation.

That said, there was at least one Tokyo radio report, heard at the American base at Guam, stating that not one but several "parachute-borne atomic bombs" had been dropped on Hiroshima. The report was picked up by the United Press wire service and created confusion about whether Washington's announcement or the enemy report was correct. The Japanese radio announcer added that "by employing a new weapon destined to massacre innocent civilians, the Americans have opened the eyes of the world to their sadistic nature."

Then, on August 15, an even more astonishing announcement was broadcast. Japan's Emperor Hirohito—considered a living god by his subjects, most of whom had never before heard him speak—informed his nation that, due to a "new and most cruel bomb" being used against the nation, Japan was surrendering to the Allies. (The surrender was billed as unconditional, but Hirohito was being permitted to remain in position as emperor—a concession previously denied by the Allied powers.) If the Japanese continued to fight, the emperor continued, not only did the country face obliteration but the conflict could quite possibly lead "to [the] total extinction of human civilization."

Celebrations erupted around the world. The Victory over Japan Day, or V-J Day, celebrations in New York City dwarfed the V-E Day celebrations of May. Two million people jammed into Times Square and the surrounding streets this time. When the *New York Times* ran the words "Official—Truman announces Japanese surrender" across its electric zipper sign on the Times Tower there, "the victory roar . . . beat upon the eardrums until it numbed the senses," recalled one *Times* correspondent. The party was "instantaneous and wild," and the "metropolis exploded its emotions with atomic force." This time the joy had a harder edge to it. Nearly a thousand people were treated for injuries incurred in the celebrations. Fourteen thousand police plus air raid wardens, more than a thousand Navy shore patrolmen, and four companies of military police were called in to suppress "over-exuberance." Some revelers grew hysterical in the streets; others sobbed openly. Thousands crowded into churches and synagogues for services. American flags hung in store windows across the city and fluttered from balconies and fire escapes and cars; once again, shredded paper swirled like smoke in the air and piled up knee-high in the streets. Sailors and Army men fanned out in the streets, grabbing and kissing girls. More than a dozen effigies of Emperor Hirohito were strung up on light poles around the city and later cut down and burned; small boys toted handwritten placards proclaiming, "HANG THE EMPEROR." The next day the delirium began all over again.

Few seemed to share Hersey's qualms and distress about the means by which the Americans had brought the war to an end at last. A poll conducted the day after V-J Day revealed that the vast majority of those surveyed approved of the nuclear attacks on Japan. Nearly a quarter of those polled in a separate August survey stated that they wished the United States had been able to use even more atomic bombs on Japan before the emperor had surrendered.

THE FIRST-INS

American leaders immediately urged the public to look ahead instead of reflecting on the war. On the evening of V-J Day, New York City mayor Fiorello La Guardia broadcast a radio speech: it was indeed a moment for joy and rejoicing, but there was a great deal of work ahead, he said. Having "defeated and destroyed forever the Nazis, the Fascists, and now the Japs," he said, "we must live up to all that this means." The tasks of reconstructing and instilling democracy in Europe, and bringing postwar order to the United States, needed to begin "within an hour or two."

Many Americans, ecstatic but exhausted, were glad to leave behind the war's horrors and focus on the future. Yet not everyone was ready to move on so quickly from the final days and acts of the war. As the days and weeks went on, there was still little information published in the mainstream American press about the aftermath in Hiroshima and Nagasaki—mostly because Western journalists had not yet been able to get into Japan. Yet the Japanese media had now begun reporting freely on the aftermath of the bombings, and disturbing reports began to filter over to the United States about lingering radiation killing off survivors of the blasts. The timing could not have been worse: U.S. forces were converging upon the Japanese islands, preparing to move tens of thousands of occupation troops into the country—including the atomic cities.

Then, on August 31, 1945—more than three weeks after the Hiroshima bombing—the *New York Times* ran an account by the first Western journalist to get into that city. Former United Press (UP) journalist Leslie Nakashima—who had before the war possessed both American and Japanese citizenship and been stranded in Japan for the duration of the conflict—had gotten into Hiroshima on August 22 to search for his Japanese mother amid the ruins. (She had been on the outskirts of the city when the bomb was dropped, and survived.) On August 27, UP, a wire service (whose name later became UPI), had run and distributed his eyewitness account of what he had seen there. The city of 300,000 had vanished, Nakashima reported. Not a single building had been left standing intact; Hiroshima was a horrific landscape of rubble and ash.

In his original UP story, Nakashima also reported that Little Boy had not finished its handiwork on August 6. Blast survivors "continu[e] to die daily from burns suffered from the bomb's ultra-violet rays," he reported, adding that "the majority of the cases [at surviving hospitals] are held to be hopeless." Many of the survivors he saw had been burned beyond recognition. Wild rumors were now circulating there about the true nature of the American bomb: that the uranium it had given off had seeped into Hiroshima's ground; that the city would be uninhabitable for the next seventy-five years; that the radiation poisoning being suffered by blast survivors came from "inhalation of the bomb's gas." Nakashima reported that he personally had "inhaled uranium" and had since been suffering from exhaustion and total loss of appetite.

Four days later, buried on its fourth page, the *New York Times* ran an abbreviated version of Nakashima's UP account—omitting nearly all references to radiation and uranium poisoning, and adding an editor's note stating that "United States scientists say the atomic bomb will not have any lingering after-effects in the devastated area." The heavily edited story now indicated that victims were dying solely of burns and injuries incurred from the blast, not radiation poisoning. Also, immediately below this story, the *Times* ran an item headlined "Japanese Reports Doubted," in which the head of the Manhattan Project, Lieutenant General Leslie Groves, was described as contending that "Japanese reports of death from radioactive effects of atomic bombing are pure propaganda."

"I think our best answer to anyone who doubts this is that we did not start the war," General Groves added, "and if they don't like the way we ended it, to remember who started it."

Yet a few days later, in early September, another harrowing initial press report emerged. Now that the U.S. occupation forces were entering Japan, scores of foreign reporters were getting in too. Several tough accredited Allied war correspondents now vied for the first major breakthrough story from the ground in Hiroshima and Nagasaki. Australian war reporter Wilfred Burchett, of London's *Daily Express*, managed to make it into Hiroshima, even though Western correspondents had been forbidden by occupation authorities to travel throughout the country. Burchett had come into Japan on a Marine-laden U.S. freighter from Okinawa and promptly boarded a train to the atomic city, which looked to him like it had been not just bombed but steamrolled. The *Daily Express* ran his findings under a banner headline proclaiming "THE ATOMIC PLAGUE."

This was his "warning to the world," Burchett wrote, about the true nature of the bomb. (What he had seen, as he later put it, was not just the end of World War II but "the fate of cities all over the world in the first hours of a World War III.") The physical devastation had been just staggering, unfathomable. Not only had the entire city been pulverized, the Japanese reports about radiation poisoning were not lies or propaganda after all. He had personally witnessed evidence to the contrary. Thirty days after the bombing, people in Hiroshima were still dying "mysteriously and horribly"—including people who had been uninjured by the blast. Their hair was falling out; they were bleeding from their ears, noses, and mouths. Helpless doctors were administering vitamin A injections, only to see their patients' flesh rot away from the injection holes; in every case, the victim died, Burchett reported. Doctors had no idea what was causing the "plague" but suspected that "it is given off by the poisonous gas still issuing from the earth soaked with radioactivity by the split uranium atom." The newspaper ran an aerial shot of the devastated city, with a caveat heading: "THE PICTURE THAT DOES NOT TELL THE WHOLE STORY."

That same day—September 5—the *New York Times* reversed course and ran its own story from the ground in Hiroshima—this time on its front page and written by Hersey's friend, Bill Lawrence. Hiroshima was indeed the world's "WORST DAMAGED CITY," the headline proclaimed. Lawrence reported that he had "never looked upon such scenes of destruction." In the air hung "the awful, sickening odor of death." He confirmed Burchett's account that blast survivors were indeed suffering from an awful affliction; the bomb had mysterious, terrifying, lingering effects after all. Symptoms included soaring fevers, dramatic hair loss, near-total loss of white blood corpuscles, and lost appetites, and most of the victims "vomited blood and finally died."

But then the *Times* and Lawrence appeared to quickly backtrack. Less than a week after his "Worst Damaged City" story was published, Lawrence had a new article out, with the section header "FOE SEEKS TO WIN SYMPATHY."

In the story, Lawrence wrote that he was now convinced that "horrible as the bomb undoubtedly is, the Japanese are exaggerating its effects . . . in an effort to win sympathy for themselves in an attempt to make the American people forget the long record of cold-blooded Japanese bestiality." It was a bewildering retreat. Something was clearly going on behind the scenes.

THE RIGHT KIND OF PUBLICITY

If Hersey had been distressed at the time of the bombings, the initial press reports out of Hiroshima only made him more uneasy and upset. Not long after Bill Lawrence's first Hiroshima report appeared in the *New York Times*, Hersey received a letter from him. Despite the horrors he had just witnessed and reported on, Lawrence was giddy over his initial aftermath scoop.

"Most of it has landed on [page one] of the New York Times, a newspaper which you may read from time to time," he bragged to Hersey. "The atomic bomb was all that everybody said it was," he went on, "except I don't think that it leaves any lingering radio activity. At least I hope not. At least I hope it doesn't make everybody sterile. At least I hope it doesn't make me that way."

Lawrence told Hersey that he had accessed Hiroshima not as an independent reporter, as Wilfred Burchett had, but rather as part of a government junket, staged by an air force press relations officer. That July, just ahead of the Japan bombings, a select group of newspaper and radio correspondents and still and newsreel photographers had been urgently summoned to the Pentagon. Lawrence was among the chosen, along with correspondents from the Associated Press, United Press, the *New York Times*, NBC, CBS, and ABC, among other outlets.

At the Pentagon, they had been greeted by Lieutenant Colonel John Reagan "Tex" McCrary, a reporter turned public relations officer for the

U.S. Army Air Force. Reporters would later recall Lieutenant Colonel McCrary—born on a Texas ranch called Wildcat Farm—as dynamic and jaunty. The consummate showman, McCrary would later become a radio and television personality, and help pioneer the morning talk show format.

Lieutenant Colonel McCrary informed the gathered reporters that they had been selected for the greatest assignment of the war. ("What, another?" one reporter scoffed.) He had been instructed by his superiors to showcase to reporters the air force's handiwork during the war, but there was, reporters were told, another story that they had been selected to cover: an "earth-shaking event which would change the course of history, [something that] was ultra-ultra secret" and would take place in the Pacific. Apparently it had been decided that *some* publicity for America's new bomb—the right sort of highly controlled publicity—was necessary. The story of the bomb's extreme destructiveness—and therefore the United States' new, powerful status as creator and sole possessor of the bomb—needed to be showcased to its allies and adversaries alike.

McCrary had not been involved in the Manhattan Project beyond having asked General Groves if he could go on the Hiroshima bombing mission. (His request was declined.) His own mission was to be more of a luxe conflict-aftermath sightseeing tour for the reporters. The McCrary junket would take place on two gleaming Boeing B-17 Flying Fortresses— which McCrary had dubbed the *Headliner* (whose name was painted in black capital letters across the plane's nose) and the *Dateliner*—which had been outfitted with plush seats, desks, lamps, and then state-of-the-art long-range radio transmitters. Above Lieutenant Colonel McCrary's desk hung a CENSORED stamp.

The junket had begun in Europe, so the reporters could survey the bomb damage in Europe's cities. They could then eventually "compare it with the damage in Japan after we defeat[ed] Japan," recalled one reporter, and therefore emphasize its comparative magnitude. The junket had just started making its way to Asia when, on August 6, they got the news of the Hiroshima bombing along with the rest of the world. By the end of the month, the *Headliner* and the *Dateliner* had converged upon Japan as part of the first wave of the press corps. Lieutenant Colonel McCrary had the planes flown over Nagasaki so the reporters could behold the devastation from above. They were encouraged to transmit their first impressions back to their news outlets right away.

"I ad-libbed my report to the *Times* into a microphone as our aircraft circled Nagasaki, and my military censor, Lieutenant Colonel Hubert Schneider, an intelligence officer based on Guam, sat close beside me to listen," Bill Lawrence recalled later. "Colonel Schneider actually assisted me in framing the report by providing military intelligence descriptions of Nagasaki's appearance before it had been hit."

McCrary's goal at Nagasaki had been to get the press to report on the atomic bombing without getting too graphic or revealing too much about the aftermath. Even when the junket was taken into devastated Hiroshima and Nagasaki a few days later, the correspondents were allowed only a few hours on the ground in each place. They were horrified by what they saw. Hiroshima was a decimated "death laboratory" littered with the corpses of "human guinea pigs," recalled one McCrary junketeer later. As they walked through rubble and ashes of the city, the McCrary reporters actually ran into Australian correspondent Wilfred Burchett of the *Daily Express*, who had then been typing his "Atomic Plague" article furiously on his portable Hermes typewriter in the middle of the smoldering ruins. Burchett was contemptuous of this group of "housetrained reporters" who were simply "being rewarded for [their] faithful rewrites of the Washington headquarters communiqués" with the promise of the greatest scoop in history: a first look at the results of America's new war-winning weapon. What they had actually been selected for, Burchett later wrote, was participation in a cover-up of outsized proportions.

When describing his junket experience in his letter to Hersey, Bill Lawrence left quite a bit out. He did not mention that when they reboarded the *Headliner* to file their Hiroshima reports, Lieutenant Colonel McCrary instructed the reporters to downplay the grotesque details of what they had seen there, as Americans were "not ready for it back home." Nor did Bill Lawrence tell Hersey in the letter that when the junketeers got back to Tokyo, General MacArthur—now Supreme Commander of the Allied Powers and effectively Japan's new emperor—and his officers were quickly clamping down on both the Japanese and foreign media. Enraged by the McCrary mission, General MacArthur was said to be threatening to court-martial the entire entourage. Despite the onboard censoring, some of those initial published reports filed by the McCrary junketeers via the *Headliner*'s transmitter—including Bill Lawrence's first article for the *New York Times*—had flagrantly crossed the line from good bomb PR into bad PR. Even worse, Wilfred Burchett's independent "Atomic Plague" story for the *Daily Express* had also just broken and was creating a worldwide outcry. (It had been a miracle that Burchett had gotten his story out in the first place: the article had had to be transmitted from Hiroshima to a Burchett colleague in Tokyo via a Morse code handset.)

U.S. government officials in both Tokyo and Washington, D.C., realized that the press and the story needed to be managed—immediately. U.S. forces in Japan quickly declared the atomic cities off-limits to reporters and corralled them into what Burchett called a "press ghetto" in Yokohama, a landing point for U.S. forces. Occupation authorities stationed sentries on the bridges over the river running between Yokohama and Tokyo. They found other ways to punish Wilfred Burchett when he returned to Tokyo from Hiroshima. Hospitalized after exhibiting what appeared to be symptoms of radiation poisoning, Burchett brought along his camera, whose film was full of images of Hiroshima's devastation; it went mysteriously missing during his hospital stay. When he emerged, he found that "General MacArthur had withdrawn my press accreditation," he recalled later. "I was to be expelled from Japan for having gone 'beyond the boundaries of "his" occupation zone without permission.'"

Now installed and getting organized in Tokyo, General MacArthur's censors had wised up fast, and a few days later they managed to shut down what would have been yet another damaging report from a third foreign reporter. A pugnacious American war correspondent named George Weller—*Time* magazine had once described him as the "much machine-gunned George Weller"—had separately made his way into Nagasaki, and had been attempting to report to his newspaper, the *Chicago Daily News*, on the devastation there.

Weller had zero respect for General MacArthur's restrictions and censors. "I had a right to be in Nagasaki, closed or not," Weller said later. "Four weeks after the two bombs, with no riots or resistance in Japan, it seemed reasonable that MacArthur should lift his snuffer from the two cities. . . . I was not going to be stifled." If Lieutenant Colonel McCrary had warned that Americans weren't ready for the truth about Hiroshima and Nagasaki, Weller believed exactly the opposite. What the United States badly needed, he thought, was "a long cold bath of reality"—not just its government, but its citizens.

Like Wilfred Burchett, Weller had managed to sneak away from General MacArthur's occupation troops and had even impersonated an American colonel to get local Japanese police to protect and assist him in his reporting. Once on the ground in Nagasaki, he had remained there for days, and during that time wrote 10,000 words describing, in graphic detail, a sinister "Disease X" ravaging blast survivors. (Ironically, like Burchett, he encountered the McCrary press junket when it dipped in and out of Nagasaki. To Weller, the reporters looked "like yacht passengers who have stopped to buy basketry on an island.") Still posing as a colonel, he had conscripted the services of Japanese Kempeitai, or military police, to transport his copy up to Tokyo for transmission back to the United States. The censors in Tokyo were apparently not as naïve as his messengers, for Weller's dispatches were apparently intercepted, rejected, and then "lost."

All of this information would come out much later. But in the mean-time, all that Bill Lawrence's letter told Hersey on September 10 was that the McCrary junket had been quite a "party," a "fabulous trip." He would be back in the States soon. At the moment, however, he was still sitting on the panel-walled *Headliner* B-17 as he wrote the missive to Hersey, enjoying a view of Mount Fuji.

"I have been having the time of my life," he said to Hersey. "Jealous, bud?"

Whether Hersey was jealous of Lawrence's Hiroshima access and coverage at the time is unknown. But even during those early weeks after the bombing, he knew that something was deeply awry in the story of Hiroshima as it was presented to the public. "As a journalist," he later recalled, he would have "no choice but to write about the world that was born [when the first bomb was dropped on Hiroshima]." It was just a matter of time.

Chapter Two

Scoop the World

THE HAYSEED AND THE HUNCH MAN

As one might expect, the *New York Times* headquarters at 229 West Forty-Third Street was an impressive affair. Times Square itself had taken its name from the newspaper's nearby original building, once the second tallest in the city. The newspaper's new building exuded power and gravitas. The limestone and terra-cotta "chateauesque" structure boasted eleven floors. And as the *Times* operation continued to grow, the newspaper's owners added floors, wings, and extensions, making it, the *Times* declared, "the greatest and completest newspaper workshop in the world."

Just up the street from the *Times* flagship stood the headquarters of quite a different publishing operation: that of the *New Yorker* magazine. Unlike the powerhouse publishing operations nearby, the *New Yorker*'s offices—spread out over a few floors of 25 West Forty-Third Street—were ramshackle by anyone's standards. "Squalor had come to be something that members of the staff took considerable pride in," said one longtime *New Yorker* contributor. "It was as if the magazine, having decided that it could not afford to be beautiful, had decided to be as ugly as possible." Plaster would occasionally fall out of the ceiling; paint peels curlicued on the walls; staff writers and editors labored in offices that resembled "bleak little ill-painted cells" lining the main corridor.

Standing in the elevator lobby was a brazier stuffed with cigarette butts and wadded-up rejection slips: many writers and artists aspired to become *New Yorker* contributors but didn't make the cut. Brilliance and a certain sort of shrewdness were de rigueur here. Nor was the magazine meant to be read by just anyone: the *New Yorker*'s founder and editor, Harold Ross, would panic when his readership exceeded 300,000. ("Too many people," he once said. "We must be doing something wrong.") From the beginning he had pledged that his magazine—originally founded as a humor publication—would cater solely to urban sophisticates. The provincial sensibilities of the "old lady in Dubuque," as Ross had put it when launching the *New Yorker* in 1925, would be diligently ignored.

Now that Hersey was free from *Time* magazine and Henry Luce, he could write for any publication. He wanted the *New Yorker*, and luckily the *New Yorker* wanted him too. Hersey knew it would probably make Luce furious when he discovered that his mishkid successor was swapping him out for the *New Yorker*: the Time Inc. boss and Harold Ross hated each other. Luce's obvious hyper-patriotic agenda repulsed Ross, as did the "*Time*style" in which Luce's editors required the magazine's correspondents to write. The style was stagy and hokey, Hersey admitted, and often featured idiosyncratic backward sentences. For the *New Yorker*'s wits, both Luce and *Time*style were perfect targets for satire—something in which Ross and his own writers specialized. ("Backward ran sentences until reeled the mind," wrote one *New Yorker* contributor in a gleeful *Time* send-up. "Where it all will end, knows God!")

Hersey had actually published his first story in the *New Yorker* a year earlier, when he was still on Luce's payroll: a profile on a young naval lieutenant named John Fitzgerald Kennedy and his recent ordeal in the Solomon Islands, in which a Japanese destroyer had charged his PT boat and cut it in half, killing two of his men. Kennedy, skipper of the PT boat, had spearheaded the rescue of his crew, personally towing the most injured men to a nearby deserted island.

Kennedy just happened to be the former paramour of Hersey's wife, Frances Ann. She and Kennedy had remained friendly after they parted ways romantically, and one evening in February 1944 the Herseys and Kennedy had met up at a chic Manhattan nightclub, where Kennedy told them his story. Right away, Hersey informed Kennedy that he wanted to do an in-depth article about his experience. ("He was then, Kennedy, son of [former] Ambassador [Joseph] Kennedy, and in that sense, he was a newsworthy figure," Hersey later recalled. But, he added, "it was quite clear that it was a good story, whether he was a Kennedy or not.") Having now accrued five years of war reporting experience, Hersey had become fascinated by and practiced in writing stories about survival and human tenacity under extreme duress. He had first brought the Kennedy profile to editors at *Life* magazine, also a Luce publication. To Hersey's surprise, they rejected it.

This ultimately turned out to be another regrettable decision on Team Luce's part, for it gave Hersey the bridge to what would become his new publication. He was given permission to shop the story around, and so brought it to William Shawn, Harold Ross's deputy editor at the *New Yorker*. Shawn immediately seized the opportunity. He had, in fact, "been trying for two years to get a story out of Hersey," Ross reported to Kennedy's father, Joseph Kennedy, and had therefore been "in a glow of satisfaction at having finally got one."

The *New Yorker* was down to a bare-bones wartime staff at that point; many writers, artists, and editors at the magazine had been drafted or had enlisted. Ross and Shawn were putting in six- and seven-day weeks, overseeing all aspects of the nonfiction stories in the magazine. (Ross complained to one writer that he was "up to [his] nipples in hot water.") It was Hersey's first experience with working with one of the most colorful odd-couple teams in the industry. Ross especially intrigued and amused him. "He had a wide countryman's mouth, his complexion was as coarse as the face of the moon, and he wore his hair on his biggish head cropped to about two inches in length, so it stood up in all directions," Hersey recalled later. It seemed wonderfully ironic to Hersey that "the editor of a sophisticated city-slicker magazine—a man who, late at night, was always given the most elite table in the back room at the Stork Club, then the smartest night spot in town—should look like a hayseed, an absolute rube."

Ross was an extravagant personality with a talent for profanity. He occasionally brandished a knitting needle as a pointer during meetings, after which writers were often dismissed with a wave of the hand and a brisk utterance: "All right . . . God bless you." He often defaced his writers' drafts with dozens of queries and edits written on the margins. (The Ross editing experience could be like "being stung to death by an army of gnats," one *New Yorker* writer recalled.) When Hersey submitted the Kennedy story to the editors at the *New Yorker*, his draft was returned to him covered with more than fifty Ross edits in the margins.

"Interrogatory howls of outrage," Hersey recalled. "I had put it toward the end of his ordeal [that] Kennedy, encountering some natives, had given them a coconut, on which he 'wrote a message.' 'With *what*, for God's sake?' Ross's note asked. 'Blood?' "

William Shawn, on the other hand, was introverted to the point of near invisibility. "I am there, but I am not there," he told his *New Yorker* writer and mistress, Lillian Ross. He was so diminutive that to some he resembled an elf. When the U.S. government made ominous noises about the possibility of drafting Shawn as well, Ross had fended off the effort, describing his deputy to a White House official as "thirty-seven, flat-footed, stoop-shouldered, a pill eater, hopeless for any service more deadly than behind a typewriter."

Shawn's demeanor could, apparently, verge on holiness. "He was extraordinarily charismatic in a paradoxical way," recalled one *New Yorker* editor. "He seemed extremely shy, extremely deferential, and yet he had this extraordinary power. People would break into tears when they sat down to talk with him. He had a . . . strange presence." His writers described his all-encompassing compassion as almost unfathomable. To him, "every human being [was] as valuable as ever other human being . . . every life was sacred," recalled Lillian Ross, who was incredulous that Shawn truly believed in the value of *every* human life.

"Even Hitler?" she asked him.

"Even Hitler," Shawn replied.

Despite their differences in temperament, Ross and Shawn were a perfect editorial team, incisive and shrewd. Both were unabashed, relentless perfectionists; both had a near mania for accuracy. In their youths, each had ditched their respective school classrooms for newsrooms.

The Second World War had brought out the deeply competitive news hound in each of them. Ross had long instructed Shawn that "we don't cover the news; we parallel the news"; yet, when the Japanese attacked Pearl Harbor on December 7, 1941, this approach went out the window. In the months and years that followed, the *New Yorker* dispatched correspondents to fronts around the world. The magazine had been born and raised to cover the "world of speakeasies, nightclubs, and chorus girls"— yet Ross had always, from the early years of the magazine, also craved gravitas. When the war came, he had a choice. He could try to maintain the original hijinks tenor of his magazine, now suddenly totally out of step with the times ("Nobody feels funny now," he complained to his ex-wife and *New Yorker* cofounder, Jane Grant), or he could take advantage of "the greatest journalistic opportunity in history," as Shawn later put it. They chose to chase the historic opportunity.

In wartime mode, William Shawn had become the "hunch man," sending his reporters to remote locations without knowing exactly what the resulting story would be—just that there would be one. And Ross had gotten quite good at getting exclusives even out of journalist-flooded, war-devastated landscapes. In 1945, *New Yorker* correspondent Janet Flanner reported from the ruins of Cologne, Germany—yet another city "pulverised by bombs"—on atrocities inflicted there by Germans upon their prisoners. Although other war correspondents had been with Flanner in Cologne, she alone got this hiding-in-plain-sight story. Her resulting series was so penetrating and effective that Ross grew giddy with competitive pride.

"How simple it is to scoop the world, even if a flock of other journalists have the same facts and the same opportunities," he told Flanner.

Yet the giddiness only went so far. Many huge, important war stories were simply going untold, Ross fretted, either out of journalistic obliviousness, a demoralizing overabundance of atrocity stories, a limited bandwidth among readers to absorb such stories, or all of the above. The war would be over and "any number of real atrocity stories" were likely to be overlooked and forgotten, he told Flanner.

"Unless the *New Yorker* gets around to doing something," he added.

For the most part, the team at the *New Yorker* carried on their native attitude of defiance. "We published what pleased us" without caring who else they pleased, claimed Shawn. That said, Ross and Shawn played by the wartime rules set up for press by the U.S. government, sometimes going above and beyond in their cooperation with the military. They dutifully submitted their stories to the War Department for mandatory censorship—including Hersey's Kennedy profile (which was quickly cleared, with only minor changes). They flatteringly profiled Allied soldiers and military personalities. The *New Yorker* was distributed to U.S. armed forces around the world in "pony" editions—small, condensed versions of the actual magazine. Ross and Shawn even published some works submitted by War Department public relations officers, likely to keep relations nice and cozy. For Ross, cultivating and maintaining access at all levels was crucial.

For Hersey, working with Ross and Shawn had been revelatory. Despite being put through the Rossian wringer, it had been a "very important experience" for him—one that had made him first realize that it was time to move on from *Time* and Luce. As an editor, Shawn had proved a writer whisperer who "had [a] genius," Hersey recalled, "for helping me find the word that I would most naturally use."

Hersey's Kennedy story, "Survival," ran in the June 17, 1944, issue. Joseph Kennedy made no effort to mask his disappointment that the story had gone to the *New Yorker*: for him, the magazine was too small, too niche—an assessment that rankled Ross, who saw his wartime magazine as a scrappy underdog competing against powerhouse news-gathering outlets.

"We have long had a feeling here that we are kicked around a great deal by the big fellows, or [on] behalf of the big fellows," he told the elder Kennedy, adding—when Kennedy pushed him to release the article to a magazine with a larger publication—that "we are not disposed to lay down now."

Joseph Kennedy ultimately harangued Ross into allowing the story to be reprinted in *Reader's Digest*, a large-circulation magazine that Ross also found time to revile. Kennedy commissioned 100,000 copies of "Survival" to be distributed throughout Massachusetts's Eleventh Congressional District during his son's 1946 run for the House of Representatives; the article would be distributed in bulk during subsequent John F. Kennedy election campaigns as well, displaying his credentials as a war hero. Hersey's reporting has been widely credited with helping to launch the thirty-fifth American president's political career. However, in first bringing Hersey to the *New Yorker*, "Survival" had ultimately played a vital role in launching two seismic careers at once.

WHAT HAPPENED TO THE HUMAN BEINGS

In late fall 1945, Hersey and Shawn had lunch. It likely wasn't a glamor-ous affair; "[Shawn's] idea of a power lunch was orange juice and cereal in the Rose Room at the Algonquin Hotel," long the regular gathering place for some of the *New Yorker*'s most vicious and brilliant contri-butors.

Hersey had written a couple of military-friendly pieces for Ross and Shawn since the Kennedy profile, including a piece on a literacy pro-gram in the Army and a story on a U.S. Army division that had returned from Europe and was being redeployed for occupation in Japan. He had decided it was time for him to begin overseas reporting again and was planning a major, monthslong trip to Asia. He would start in the country of his birth, China, to report on the aftermath of the war there and then try to get into occupied Japan.

At lunch, as Hersey and Shawn went through possible story ideas, they discussed Hiroshima. The *New York Times*, the only news outlet to have an eyewitness reporter on the Nagasaki bombing run, would boast to advertisers that it had scooped the world on the story of the atomic bomb and the dawn of the atomic age; it remained, as one *New Yorker* writer put, "the biggest news story in the history of the world." But for Hersey and Shawn, something essential was missing from the reporting. They identified what had seemed so disturbing and incomplete about the coverage so far.

"Most of the reporting up to that time had to do with the power of the bomb and how much damage it had done in the city," Hersey later re-called. While it seemed like the coverage had been comprehensive, most of the information had dealt with landscape and building destruction. Months had now passed since the nuclear bombing of Hiroshima, and still very little had been published on how the atomic bomb had affected its human victims. Scores of seasoned, ambitious foreign correspondents were now based in Tokyo—and yet no one had yet followed up those alarming first dispatches from Hiroshima with a major, comprehensive story on the fallout there.

Famed *Life* photographers Alfred Eisenstaedt and J. R. Eyerman had by then both traveled to and photographed Hiroshima, but the images published in their magazine were fairly sanitized. (Eisenstaedt took one now-published heartbreaking portrait of a Japanese mother and toddler amidst the charred ruins, but it never ran in *Life*'s national edition.) A spread of Eyerman's photos did include two rare pictures of Japanese A-bomb victims, yet their wounds appeared negligible; the caption on one picture reminded readers that their burns evoked those received by American soldiers at Pearl Harbor. Other published Eyerman pictures featured little Japanese girls sporting sun hats and parasols; another featured Japanese soldiers reclining around what appeared to be a relatively intact train station. The message: post-bomb life in Hiroshima wasn't so apocalyptic after all.

Furthermore, what limited coverage there had been was now tapering off. Other stories had since edged out front-page Hiroshima and Nagasaki headlines. When Americans opened their newspapers each morning, they instead saw news of the homecomings of American troops, the reconstruction of Europe, the Nuremberg trials of war criminals in Germany, and, of course, the escalating antagonism between the United States and the Soviet Union, among other international developments.

It is unclear whether Hersey and Shawn knew the extent of the restrictions that had been imposed on Tokyo-based reporters, but they likely had some idea. The American journalism world was then very close-knit, and many of Hersey's correspondent friends had been posted to Japan to cover the occupation. He knew Eisenstaedt from Time Inc., for instance; the photographer had taken Hersey's photograph during the height of the author's *A Bell for Adano* fame.

In any case, it was clear that the U.S. government had been put on the defensive after the initial Hiroshima press reports came out, during the chaotic first days of the occupation. Both in Washington, D.C., and in General MacArthur's Tokyo, U.S. officials had immediately gone into overdrive to contain and spin the story of the A-bomb aftermath.

Right after the Tex McCrary junket reports and Wilfred Burchett's "Atomic Plague" story came out, editors of American publications—Ross and Shawn likely among them—had received a confidential September 14 letter from the War Department, on behalf of President Truman, asking them to restrict information in their publications about the atom bomb. The War Department was quick to state that this was not an extension of wartime censorship, which was officially slated to end that fall. Rather, the War Department directive said, it was simply a request that newspaper and magazine editors submit any material relating to nuclear matters to the War Department for review. It was an issue of "the highest national security" lest proprietary information about the bomb fall into foreign hands.

Starting around that time, there had also been a hastily assembled offense of government press conferences meant to set the press straight and the American conscience at ease. Even before the occupation had officially begun, Charles Ross, press secretary to President Truman, had sent the War Department a memo suggesting a media event at the site of the July 16 bomb test in New Mexico to show the public that there was no long-term radiation damage there, and absolutely nothing to worry about when it came to post-bomb aftermath. "This might be a good thing to do in view of continuing propaganda from Japan," he wrote.

On September 9, Manhattan Project leaders General Leslie Groves and J. Robert Oppenheimer personally gave a tour of the New Mexico bomb testing site to a group of around thirty journalists. Even though the reporters were assured that there was barely any residual surface radiation, they all wore special white protective shoe covers, just to "make certain that some of the radioactive material still present in the ground might not stick to our soles," one reported.

"The Japanese claim that people died from radiations," General Groves told the reporters. "If this is true, the number was very small."

The attending journalists toed the line. One *New York Times* correspondent, William Laurence—who also happened to be on the War Department payroll and had been "on loan from the *Times* to the atomic bomb project" since the previous April, as the *Times* put it in personnel records—reported that "the Japanese are still continuing their propaganda aimed at creating the impression that we won the war unfairly, and thus attempting to create sympathy for themselves and milder terms." In the meantime, he went on, the New Mexico ground on which he had personally stood "gave mute testimony" that the "Tokyo Tales" of death by radiation were indeed fiction. Their Geiger counters had revealed that "radiations on the surface had dwindled to a minute quantity."

The U.S. government and the Manhattan Project principals actually had not known in advance what the full effects of their experimental weapons would be, and they now hustled to make private investigations in the "death laboratories" of Hiroshima and Nagasaki. They urgently needed to find out whether radioactivity was indeed lingering in the atomic cities and how it was affecting humans—not because the United States intended to help treat Japanese bombing victims, but rather because those cities were about to be occupied by American troops.

Just before giving his nothing-to-see-here New Mexico press event that September, General Groves had hurriedly dispatched Brigadier General Thomas F. Farrell, his Manhattan Project deputy, to Japan to survey the aftermath in Hiroshima. On September 8, General Farrell and a team of Manhattan Project scientists arrived in the city to make what one reporter later described as a "spot check" inspection. (Upon seeing the damage in Hiroshima, one attending American physicist thought that, to the average beholder, it looked like the handiwork of a thousand B-29s, but that the United States had instead "simply used a labor-saving device.") From Washington, the pressure on General

Farrell was intense, recalled a member of his team; General Groves continuously bombarded Farrell with indignant cables from afar, demanding updates.

In Hiroshima, the team was able to calculate the height at which the bomb had detonated. "The bomb had burst at precisely the spot we wanted it to, high over Hiroshima," recalled one Manhattan Project physicist. They therefore concluded that "there had been a minimum of radioactivity in the city," as most of it had been absorbed back into the atmosphere. Based on the Farrell team's brief investigation, General Groves announced to the press that very few Japanese had actually died from exposure to radiation and that Hiroshima was essentially radiation-free.

"You could live there forever," he said.

Once back in Tokyo, General Farrell held his own press conference at the Imperial Hotel to present his atomic city findings. He did concede that a few Japanese were now dying due to gamma ray exposure received from the blast on August 6, but that the reporting on Hiroshima had been exaggerated. The conference was going well until the *Daily Express*'s Wilfred Burchett materialized, having just arrived from his own "Atomic Plague" reporting trip to Hiroshima. He was dirty, exhausted, and sick. When he confronted the briefing officer, in front of a full room of reporters, about what he had seen in Hiroshima, he was told that "those I had seen in the hospital were victims of blast and burn, normal after any big explosion." When Burchett pushed him further on the strange illnesses now befalling blast survivors, he was told, "I'm afraid you've fallen victim to Japanese propaganda." (Hersey's friend Bill Lawrence was present at the conference too. In his subsequent story for the *New York Times*, he did not report the dramatic Burchett exchange, and the headline of his story read, "No Radioactivity in Hiroshima Ruin." Much of the story instead directed attention away from the victims and back to the physical destruction: 68,000 buildings had been destroyed, he reported.)

In the months that followed these early-autumn press conferences, General Groves continued his own PR campaign, an all-out effort to re-brand his nuclear creation as a merciful weapon. Summoned by Congress that November to testify about use of the bombs on Japan and their after-effects, the general eventually conceded that radiation had been responsible for some of the deaths in the atomic cities, but informed the Senate Special Committee on Atomic Energy that doctors had assured him that radiation poisoning "is a very pleasant way to die." During fall visits to Manhattan Project labs and industrial contractors, General Groves gave speeches in which he told his audiences that there was no need for guilty consciences over the atomic bombings. He personally had no qualms, he told audiences.

"It is not an inhuman weapon," he told one audience. "I have no apologies for its use."

For Hersey and Shawn, it was time to get behind the scenes. The government junkets, conferences, speeches, and suppressed reporting were having the desired effect: across the United States, protests and alarm had been subdued to a manageable murmur. The idea of the atomic bomb as a reasonable mainstay weapon in the national arsenal—and a nuclear future in general—was becoming acceptable to the increasingly apathetic public. Many if not most Americans had moved on from Hiroshima and Nagasaki before the stories about what had happened in these cities had ever been told in full. To the *New Yorker* team, the true meaning of what had happened in Hiroshima—and the awful implications of those atomic bombings for the world—clearly had not sunk in.

Hersey and Shawn decided that Hersey would try to get into Japan and write about "what happened not to buildings but to human beings." They didn't know the exact angle yet, but they knew it had to be done. If the press corps in Tokyo either would not or could not undertake the story, the *New Yorker* would try. Here was one fewer wartime atrocity story that would not go ignored, as Ross had feared. Ross and Shawn had already been prepared to run another such story earlier that year, when *New Yorker* reporter Joel Sayre crossed into Germany with Allied troops and planned to do a story on the decimation of Cologne from the point of view of the civilians living there during the bombing campaign. That story did not come to fruition—but its approach could be used for a Hiroshima investigation.

It was time to learn and reveal the reality of the bombs that had been detonated in the name of the country, democracy, and decency itself. It was also time to put individual faces and fates to the Hiroshima and Nagasaki casualties at last. Sterile statistics and interchangeable photographs of rubble had become the enemy of such truth telling; dehumanization of the Japanese victims was allowing Americans to be lulled into complacency about the weapon.

Throughout World War II, Hersey had seen how dehumanization had enabled the worst acts of the war. He had personally witnessed the "wanton savagery" of air attacks on civilians in Rome and evidence of Nazi atrocities at concentration camps. During the war, he, too, had seen the Japanese as an animalistic enemy, but even while under Japanese fire at Guadalcanal he had tried to remember that it was a man shooting at him—a son, possibly a husband, maybe a father.

"I had long hated the idea [of the Japanese]," he wrote shortly after the experience. The *idea* of the enemy was infinitely hateable, he stated. "But I did not hate this individual. Was he from Hakone, perhaps Hokkaido? What food was in his knapsack? What private hopes had his conscription snatched from him?"

The war—culminating in the bombings of Hiroshima and Nagasaki— had revealed to Hersey the "depravity of man" and his ability to completely degrade fellow human beings. He had come to realize that "[if] civilization was to mean anything, we had to acknowledge the humanity of even our misled and murderous enemies."

He began to plan his trip.

THE TROJAN HORSE

For Hersey and the *New Yorker* team, there was likely never any question of Hersey trying to sneak into Japan to make his Hiroshima investigation. Even Wilfred Burchett and George Weller had had to accompany Allied military forces into the country as accredited war correspondents. Of course, those reporters had been able to take advantage of the chaos of the early days of the occupation to break loose from their respective military units, but even then the control of General MacArthur (Supreme Commander for the Allied Powers, or SCAP, an acronym which also referred to his occupation administration) over the foreign war correspondents had been stringent enough that both Burchett and Weller had to conjure up elaborate schemes to get away from their public relations officer (PRO) minders. Burchett had been slated to cover, along with most of the rest of the accredited foreign press corps, the Japanese surrender ceremony aboard the USS *Missouri* on September 2; when his PROs came to collect him for the event, he pretended to be crippled with diarrhea and was left behind to "recover." He made it to Hiroshima, via train, only with the help of a sympathetic editor from the Japanese press agency Domei and a Hiroshima-based Domei fixer—which were, a year later, completely under SCAP's control and restrictions. Even though Burchett had succeeded in getting his "Atomic Plague" story out, his camera film had been apparently confiscated and his accreditation revoked.

George Weller had given his PRO the slip by sneaking away from his military base under cover of darkness; he then escaped the base in a small motorboat and taken no fewer than four trains to Nagasaki—all while impersonating an American colonel to those Japanese he encountered. Even though Weller, too, got his story, SCAP's censorship operation back in Tokyo was already airtight by the time he tried to transmit his copy back to the United States.

If General Groves had been scrambling back in the States to contain the Hiroshima story, these journalists felt from the beginning that General MacArthur would likely have an additional personal reason for vigorously suppressing reporting on the bombed cities and downplaying the bomb in general: "[jealousy] of the fact that 'his war' of four years had been won by two bombs prepared without his knowledge and dropped without his command," as George Weller put it. In his opinion, General MacArthur had "determined to do his best to erase from history—or at least blur as well as censorship could—the important human lessons of radiation's effect on civilian populations."

Now, months later, SCAP wielded near-total control over who could enter Japan; anyone wishing to enter had to apply to General MacArthur's forces for admission clearance. SCAP kept close tabs on all journalists to whom it granted entry: its Tokyo press office kept records on reporters' whereabouts, their political dispositions, their attitudes toward the occupation, and even their health. Nearly every day General MacArthur received briefings about the current press corps, with summaries of articles—favorable or critical—pertaining to the Pacific theater. There were also monthly summaries compiled on the general activities of publications, including Hersey's former employer, Time Inc. Once admitted to Japan, journalists had to be given permission to travel throughout the country as well, and they were allotted only limited time on their permissible forays.

Occupation troops had been stationed throughout the country. That past fall, around 27,000 occupation troops had moved into Nagasaki, and approximately 40,000 into Hiroshima. Most of the Hiroshima troops were housed in camps on the city's outskirts, but in Nagasaki a larger number had set up camp in the city, even in buildings near ground zero.

None of this portended an easy reporting landscape for Hersey. But for him and the *New Yorker* editors, there was one snippet of promising news: SCAP was apparently now letting some reporters make trips to

Hiroshima and Nagasaki, as shown by the *Life* magazine features. At least Hersey would have a shot at accessing Hiroshima—that is, if he was admitted by SCAP to Japan in the first place. General MacArthur detested some correspondents and hated certain publications. (The *New York Herald Tribune*, the *Chicago Sun*, and the *San Francisco Chronicle* embodied "downright quackery and dishonesty," in the general's opinion.)

For the most part, Hersey's wartime record made him a viable candidate for clearance—and, from the *New Yorker*'s point of view, the perfect Trojan horse reporter. He had played by the rules throughout the war; his body of work largely belied a dutifully patriotic reporter. (George Weller had, by comparison, been butting heads with General MacArthur's PROs and raging against the general's "iron curtain of censorship" for years.) Hersey was also a commended war hero. He had written glowing profiles on the military's rank and file. His second book, *Men on Bataan*, had been a flattering portrait of General MacArthur himself, created in close consultation with and approved by the War Department. (Hersey would later wish he could take *Men on Bataan* out of print, calling it "too adulatory" of General MacArthur, but the book was certain to be an asset in Hersey's bid to gain access to the general's Pacific kingdom.)

There was the chance that Hersey's tenure as a *Time* reporter in Moscow might prompt questions about his political loyalties, raising the proverbial red flag for SCAP officials. Another possible sticking point: Hersey's third book, *A Bell for Adano*—which contained a veiled portrait of U.S. general George S. Patton as an unhinged, destructive megalomaniac—showed that Hersey was not above rebuking the United States when he believed it fell short of the ideals for which the country professed to stand, and a willingness to show that America was not morally infallible after all. (General Patton had become "a cruel and despotic officer" who had "lost his marbles in Sicily," Hersey later stated. "Seemed to me being just the thing we were fighting our war against.")

All in all, however, Hersey's chances appeared good. Before Christmas 1945, Hersey took a transport ship across the Pacific and arrived in Shanghai. On December 30, he sent Shawn a cable: he had successfully gotten accredited to U.S. Naval Group China. This was an encouraging sign. The Trojan horse had arrived at the gate.

Chapter Three

MacArthur's Closed Kingdom

BIKINI

Despite his noble goals with the Hiroshima story, Hersey had, during the war, harbored his own prejudices against the Japanese. They were an "animal adversary," he stated in one wartime article for *Life*; they were "stunted physically," he wrote in another. Alongside another Hersey article ran a photograph of him wearing some helmet netting lifted off a dead Japanese soldier. In his book *Into the Valley*, which detailed his ordeal covering the bloody and disastrous Guadalcanal skirmish, he referred to the Japanese fanning across the island as "a swarm of intelligent little animals."

He was certainly not the only U.S. reporter to see the Japanese this way. One correspondent, Russell Brines—an Associated Press reporter who had arrived with General MacArthur's troops the previous August and stayed on as the AP's Tokyo bureau chief—fretted that "maybe the Japanese aren't human" when first entering the country, a factor that might make them unpredictable and dangerous under occupation. (When considering human reactions to radioactivity in Hiroshima, General Groves privately wondered whether there was a "difference between Japanese blood and others" that was making Japanese atomic city residents weirdly susceptible to the bomb's long-term effects.)

Hersey later admitted that he was ashamed of the way he had portrayed the Japanese in some of his wartime dispatches and writings. "Like other Americans, I had reacted badly to the humiliations of Pearl Harbor and Bataan," he said. In 1937 he had been "horrified . . . by accounts and pictures of the orgiastic Japanese rape of Nanking." In 1939, Hersey had gone to China to report on the war there for *Time*. "I had seen with my own eyes the arrogance and cruelty of the Japanese occupation of the city where I was born," he recalled. At the time, China was under relentless bombardment by the Japanese, who, it seemed to Hersey, had no strategy other than indiscriminately killing as many Chinese as possible. During his visit to Chungking, the Chinese capital, the Japanese air raids on the city "started huge, uncontrollable fires, and thousands burned to death after every attack." Millions of Chinese ultimately died in the war.

Now, in 1946, Hersey was back in China. Over the next few months he would use Shanghai as his base of operations but travel to several cities and regions, from Peiping (Beijing) to Manchuria. He filed a series of stories to the *New Yorker* editors about China in the immediate aftermath of the war and the emerging battle between that country's government and Chinese Communists. He also posited to Shawn the possibility of reporting from Nanking, which would have made a demoralizing bookend to the post-atrocity coverage he was about to do in Hiroshima. This story didn't happen, but, regardless, Shawn was finding Hersey's China work brilliant. "You are doing wonders out there," he cabled the reporter.

At one point, in late March, Hersey feared that his Hiroshima story might be in jeopardy. The Joint Army-Navy Task Force Number One was in the process of planning more atomic bomb tests, this time at Bikini Atoll in the Marshall Islands. Members of the press were being invited to witness atomic detonations. The operation's declared purpose was to investigate the effect of nuclear weapons on naval warships. The experiments were to include the explosion of an atomic bomb over an assortment of target ships and boats, and another to be detonated ninety feet underwater. There was another reason behind the tests, some of the attending journalists privately concluded: "We had a monopoly on the Bomb [and] were hardly adverse to demonstrating what it could do," recalled one reporter, who pointed out that representatives of nations from around the world—including media from the Soviet Union, the United States' emerging adversary—were also invited to attend.

Hersey cabled Shawn from Shanghai, concerned that the tests would detract from the story they had envisioned. The *New Yorker*'s editors were among those approached by the task force about sending a correspondent to Bikini, with government transport available, but Shawn coolly declined the offer and instructed Hersey to proceed to Hiroshima as planned.

"The more time that passes," he informed Hersey, "the more convinced we are that [the] piece has wonderful possibilities. No one has even touched it."

If anything, the Bikini tests would ultimately make Hersey's Hiroshima assignment more urgent. The first bomb was successfully exploded over its target fleet of over ninety ships, sinking five. Dozens of journalists and photographers observed the event from a War Department ship, where there was "mirthless jesting about the possible effect of the Bikini bombs on visiting genitals." Watching the explosions from fourteen miles away, wearing welder's glasses to protect their eyes, some of the journalists found the bombing anticlimactic. Journalist Norman Cousins of the *Saturday Review of Literature* was on board. One of his fellow correspondents turned to him, unimpressed by the spectacle.

"I was just thinking," he said, "that the next war's not going to be so bad after all."

Another journalist scoffed that "the sound of the bomb . . . was like the sound of a discreet belch at the other end of the bar." There had been no chilled spines or curdled blood at all, reported another.

The newsmen aboard "recorded their petulant resentment at being cheated," recalled Clark Lee of the International News Service. "They underwrote the atom bomb's results . . . and the outcome was that many of the public decided that the reports of atomic-bomb destruction at Hiroshima and Nagasaki had been exaggerated."

One of the few journalists who discerned the true significance of the Bikini tests was William "Atomic Bill" Laurence, the veteran science correspondent for the *New York Times* who had, in spring 1945, been conscripted by General Groves to be the so-called official historian of the secret Manhattan Project. (McCrary junketeer and Hersey's friend Bill Lawrence had been given the nickname "Non-Atomic Bill" to differentiate the two *Times* reporters.) Atomic Bill—a "small, dark man . . . with a flattened nose and a wild shock of hair," as one reporter put it— had effectively functioned as a General Groves propagandist, helping to create the War Department's official statements announcing the Japan bombings while on loan to the government from the *Times*. That previous fall, Atomic Bill had informed *Times* readers that the Japanese had been exaggerating reports of bomb-related radiation; their reports, he wrote, had been just a "foe's propaganda at work."

Unlike many of the other Bikini junketeers, Atomic Bill now recognized that the Bikini tests showed that the United States' nuclear prowess was advancing quickly and that the weapons were becoming even more dangerous. In an article that bordered on alarmist—especially for Atomic Bill, the ultimate Groves lapdog—he reported in the *Times* that the second bomb detonated at Bikini had actually been the "greatest explosion ever felt on earth," this time packing a probable payload of at least 50,000 tons of TNT, some two and a half times the power of the bomb that had devastated Hiroshima. This time there *had* been lingering radiation, he stated: the tests had left behind a "churning mass of radioactive water" in the lagoon surrounded by the atolls. The spray kicked up by the blast doused all of the surrounding ships with radioactivity.

Yet many Americans were as nonplussed, even disappointed, by the tests as some of the Bikini reporters had been. The tests did less damage than expected, declared 53 percent of Americans surveyed in a subsequent poll. It was just further evidence, they thought, that the initial hysteria about the Japan bombs had been unwarranted.

THE SICKBED EPIPHANY

Hersey was determined to get into Japan by the second week of May. He applied to SCAP in Tokyo for permission to enter the country. Even though Japan remained on strict lockdown, there had been at least one positive development for Hersey while he was in China: by April 1946 the United States had largely moved military personnel out of Hiroshima, which would hopefully make it a little easier for him to maneuver on the ground. However, there were still American military police stationed there, so he would have to proceed cleverly and cautiously if he hoped to achieve his reporting mission in the city.

Toward the end of April, Hersey came down with the flu while on a side assignment to Manchuria. He was on board one of six U.S. Landing Ships, Tank (or LSTs) that were "helping to give the New 22nd Division of China's new Sixth Army a 'lift' from Shanghai to north China" to fight the Communists there—an operation overseen by the U.S. military. When Hersey got sick, he was transferred to a destroyer back to Shanghai.

It turned out to be a fortuitous flu. As he convalesced onboard, crew members brought Hersey a few books from the ship's library, including *The Bridge of San Luis Rey* by Thornton Wilder. The 1927 book detailed the lives of five people in Peru who were killed when a rope suspension bridge over a canyon broke with all five on it. Wilder's account tracked the lead-up to the accident and how these protagonists all found their way to that tragic moment.

As Hersey read feverishly in his bunk, he realized that this would be an effective way to tackle Hiroshima as a subject. Back in New York City, he and William Shawn had agreed that his goal was telling the story from the victims' point of view, but Wilder's work had just given Hersey an especially poignant, sharp way of telling the "terribly complicated story" and making it engrossing. Once on the ground in Hiroshima, he would try to track down a small number of victims whose paths crossed each other's as they came to their own "moment of shared disaster."

It was an ingenious, subversive approach. Too often, reporters had fallen back on grandiose pseudo-biblical language to describe the power of the bombs. Atomic Bill Laurence of the *New York Times* was especially guilty of this: he had just described one of the Bikini tests' mushroom cloud as a continent made "when the earth was young"; it had then, he wrote, mutated into a "a giant tree, a tree with many branches bearing many invisible fruits—alpha particles, beta rays, neutrons—fruits deadly to man, invisible to the eye, the fruits of the tree of knowledge, which man must eat at his peril." If anything, descriptions like this one had made the atomic bombs and their fallout seem even more abstract to readers. Even Atomic Bill conceded that he and others had failed to describe effectively the hugeness and significance of the bomb.

"The human mind," he decided, "is simply not conditioned to think in such dimensions."

It was time for someone to describe the bomb in terms that the human mind *could* grasp. As Hersey finished *The Bridge of San Luis*

Rey, he realized that emphasizing minutiae, not grandeur, was the way to drive the point home. Not everyone could comprehend how the atomic bomb worked or visualize an all-out, end-of-days nuclear world war. But practically anyone could comprehend a story about a handful of regular people—mothers, fathers, grade school children, doctors, clerks—going about their daily routines when catastrophe struck. Hersey would take readers into the victims' kitchens, on their streetcar commutes, into their offices, back on that sunny summer morning of August 6, 1945, and show what befell them.

"My hope was that the reader would be able to become the characters enough to suffer some of the pain, some of the disaster," he said later.

It was simply a question of scale. Hersey would dial it down from God's eye level to a human vantage point.

OCCUPATIONAIRES

On May 13, Hersey received a cable from the SCAP press office in Tokyo:

NO OBJECTION ON ENTRANCE OF
JOHN HERSEY AS CORRESPONDENT

He was in. Hersey wired Shawn from the Shanghai press office, informing the editor that he was now just waiting on military transport into Japan. A week later Hersey got another Army cable informing him that he was "hereby invited and authorized to proceed via first available transportation" from Shanghai to Tokyo.

On May 22, Hersey boarded a U.S. military plane out of Shanghai. His plane flew out over the Pacific and landed on Guam, where he boarded a naval air flight to Tokyo. Just over a year earlier, his friend Non-Atomic Bill Lawrence had circled the city in a B-29 while covering a 350-plane firebomb attack on Tokyo by U.S. forces. Japanese searchlights swept across the sky, the bright beams of light cutting through the heavy smoke hanging over the capital. The crew on Lawrence's plane dropped napalm bombs on the city below. "The incendiaries dropped by the airplanes ahead of us looked like millions of fireflies as they neared the ground," he recalled later. Within seconds, huge swaths of Tokyo had burst into flames below. The fire spread to structures on the palace grounds of Emperor Hirohito.

A year later the flames had long since been extinguished but miles upon miles of charred carnage remained. Hersey had last been in Tokyo for *Life* magazine in 1940, profiling Ambassador Joseph Grew, prewar American emissary to Japan. The city had then been a vibrant metropolis, densely packed with traditional wooden Japanese buildings—perfect kindling for the Allies' firebombs.

"In Tokyo streets, hawkers [sung] their wares," Hersey reported. "Japanese jazz, a curious marriage of Western orchestration and Eastern harmonics, [could] be heard everywhere blaring from shop doors."

Most of those shops had likely since burned and Tokyo's streets were still in shambles. The city still resembled "an ashtray filled with the cigarette butts of buildings," as reporter George Weller had described it. Twisted steel building frames jutted up from the earth; in other places, only gaping foundations marked where structures had once stood. Japanese survivors dwelled in a sea of canvas tents and crude metal-walled shacks erected on top of the rubble. Everywhere hung the smell of burning charcoal, frying fish, and human feces.

The occupation was now in full swing. Yokohama, the nearby naval port and an entry point for occupation troops, had become a garrison town filled with Quonset huts and military installations; the road between this city and Tokyo was choked with Army trucks and jeeps driving alongside Japanese handcarts, bicycles, and oxcarts. In Tokyo, U.S. troops—or "occupationaires," as they called themselves—had moved into former

Imperial Japanese Army barracks or surviving downtown office buildings. The capital's streets "flowed with unrelieved khaki and olive drab [uniforms]," recalled AP reporter Russell Brines. "The air was filled with nervous, impermanent intensity."

In Tokyo, General MacArthur had established his general headquarters in the fortresslike Dai-ichi Life Insurance Company building, right across the street from the Imperial Palace, where the Japanese emperor still lived. It was not a subtle statement. The supreme commander lived nearby in the American embassy, guarded by American soldiers "and so many Japanese police they seemed to sleep in the trees," recalled Brines. One occupationaire, an American doctor, recalled that seeing MacArthur's motorcade wind through the city was like beholding a modern Caesar. The Japanese people had been adequately subjugated, he also observed.

"When I walk down the street, my five-feet-ten-inches puts me head high over Japs who more or less crouch as they walk," the doctor commented. "The Japs are said to be a proud people, but they certainly behave like humble creatures now."

Hersey, who stood over six feet tall, was especially conspicuous in postwar Japan. (He was, noted one Japanese doctor, "one of the tallest men I have ever seen anywhere.") If he had intended to keep a low profile once he arrived, that plan was blown. His presence was immediately noted and even announced in local media. The *Nippon Times* reported that "one of America's outstanding man-of-letters" had arrived in Tokyo. He was, the story noted breathlessly, "tall, handsome, [and] youthful looking," as well as modest and polite.

As for the purpose of his visit: "[A]fter having spent several weeks in China on assignment," the *Nippon Times* reported, the Pulitzer Prize–winning author and journalist was apparently just "passing through Tokyo en route to the States."

AN INROAD PRESENTS ITSELF

Five months earlier, in January 1946, an Army filmmaker named Lieutenant Herbert Sussan had been given a daunting assignment. During the war he had been based in Culver City, California, working on propaganda films and radio broadcasts for the U.S. Army Air Corps' First Motion Picture Unit. After the war ended, he was reassigned to Tokyo, and found himself part of a small unit of U.S. military filmmakers headed by Lieutenant Colonel Daniel A. McGovern, whose mission was to document for the U.S. government the effects of American bombing on more than twenty Japanese cities, including Hiroshima and Nagasaki. When the team entered Nagasaki, Lieutenant Sussan was shocked.

"We knew that Nagasaki had been hit pretty bad but nobody prepared us before we went in," he recalled later. "I could not believe what one bomb, what one little bomb could do. These huge factories—it was as if an enormous hand had reached down from the sky and pushed them all away."

Hiroshima had been even more difficult for him to behold: its rivers and bridges reminded him of New York City, his hometown. His unit took 90,000 feet of raw footage of the atomic cities on color film stock, including disturbing shots of burned, perishing blast survivors.

In late May, Lieutenant Sussan was in the field and got a message from Lieutenant Colonel McGovern instructing him to get transport back to Tokyo for a meeting. Lieutenant Sussan dutifully materialized for the meeting at the new Tokyo Correspondents Club, founded by foreign journalists the previous fall. After having eventually been released from their Yokohama "press ghetto" by General MacArthur's forces, the fleet of journalists had made their way back into the capital to start reporting on the occupation. SCAP officials had taken up all the rooms in the remaining hotels and buildings, so the press corps found and rented a battered but habitable five-story redbrick building as their own general headquarters. It was filthy, its windows were broken, and it stood in a narrow lane surrounded by roofless, derelict office buildings, but it became home to nearly every Western journalist assigned to Tokyo. If the rooms were crammed to capacity, correspondents could sleep in the ramshackle "ball room" behind paper screens. At one point an enterprising reporter set up a little gun shop in the lobby; he divided his time between overseeing this stall and an on-site dice table. If the club was in equal parts "a makeshift bordello, inefficient gaming-house, and a black market center," as one correspondent put it, it was also a crucial hub for networking with SCAP officers, a meeting place where leaks and tips were quietly given.

When Lieutenant Sussan arrived at the club that day in late May, Lieutenant Colonel McGovern met him there and introduced him to John Hersey.

"[He's a] writer from New York who wants to do something on Hiroshima," Lieutenant Colonel McGovern told Lieutenant Sussan.

The three men sat down for lunch together. Over the meal, Lieutenant Sussan later recalled, he and Lieutenant Colonel McGovern told Hersey about their time in the atomic cities. The filmmakers' Hiroshima footage had been sent back to Washington, D.C., and would be classified for decades; Lieutenant Sussan had been told by the head of the U.S. Strategic Bombing Survey that the film was to be "limited strictly to government use." (In Lieutenant Colonel McGovern's opinion, the U.S. government "didn't want that material out because they were sorry for their sins.") Yet Lieutenant Sussan was eager to spread the word about what they had seen and filmed.

"If people could only see this devastation, this holocaust, it would be the greatest argument for peace the world has ever seen," Lieutenant Sussan later stated. The full extent of the story "had to be told . . . I remember thinking, 'This kind of weapon cannot exist on earth and people live on earth too.' "

He had personally pushed to take a good deal footage of atomic bomb victims rather than just landscape devastation footage, in order to have a record of the bombs' full effects on humans.

The filmmakers gave Hersey some Hiroshima contacts; they also told him about a group of Hiroshima-based priests who had witnessed and survived the blast and were still living in the flattened city.

Hersey had already read a report written by a Japan-based German Jesuit priest named Father Johannes Siemes, who had been on the outskirts of Hiroshima on the day of the bombing and had written an eyewitness account for a magazine called *Jesuit Missions*. In February 1946, *Time* magazine had run an abbreviated version of the translated account in its Pacific pony edition, distributed to U.S. military troops throughout that theater. Hersey also obtained a longer version of the testimony.

In this account, Father Siemes had recounted in graphic detail what it had been like when Little Boy exploded over Hiroshima at 8:15 a.m.—"The whole valley is filled by a garish light . . . I am sprayed with fragments of glass"—and the horrible scenes that had followed. After making his way through the burning city in an attempt to rescue some of his fellow priests, Father Siemes found two of his colleagues, both gravely injured. (The Japanese secretary of their mission had also survived, Father Siemes wrote, but went mad in the aftermath and apparently committed suicide by running into the flames of the inferno then consuming the city center.) The two injured priests had taken refuge in the city's Asano Park, where Father Siemes found them; they might have died there if not for the efforts of a Japanese Protestant minister—"our rescuing angel"—who materialized and helped evacuate them in a boat.

It is unclear whether Lieutenant Colonel McGovern and Lieutenant Sussan had crossed paths with the specific survivors from Father Siemes account, but at least Hersey now knew that members of this community of eyewitnesses were still alive and in Hiroshima. These priests, Hersey felt, could be his inroad to the story.

THE ARMY CONTROLS THE FOOD HERE

The Tokyo Correspondents Club was within walking distance of SCAP's press relations office. The office had been set up in the former Radio Tokyo building, near General MacArthur's headquarters at the Dai-ichi Insurance Company building. The wartime camouflage coat of jet-black paint still covered the six-story structure; inside, its rooms and hallways reeked of fish, recalled one American correspondent with disgust. On the second floor, a large newsroom with offices and desks had been set up for reporters from news organizations accredited to SCAP. Hersey was told that he could wire communications and copy back to his home offices from the Radio Tokyo building. It was a ritual that always took place under the beady eyes of General MacArthur's public relations officers, whose primary duty was to shield MacArthur and the occupation from criticism. To that end, William Shawn had instructed Hersey that he should write the story back in the States.

Hersey now had to apply to SCAP's general headquarters for permission to travel to the southern part of the country. While SCAP had had "no objection" to Hersey's entrance to Japan, he would have to maintain a diplomatic demeanor if he was going to navigate around the occupation apparatus to get access to Hiroshima. The Army kept "a tight grip on all living under it," reported AP Tokyo bureau chief Russell Brines. Everyone under its jurisdiction, Americans and Japanese alike, had even the smallest details of their lives regulated: "They were told how much to eat, how much gasoline to use and how many cigarettes to smoke."

Coverage that displeased SCAP's public relations officers could result in revoked accreditation or expulsion from the country, but the press officers had also been finding additional, unsubtle ways to keep the press under control. That past fall, General MacArthur's head of public relations, Brigadier General LeGrande A. Diller—known to correspondents as "Killer" Diller—had ensured that reporters attempting to cover a meeting between Emperor Hirohito and General MacArthur were met with bayonets, and then barred reports from being published about the incident. Pressed about his motives, General Diller replied, "Call it whimsy if you like." Afterward he informed the press corps that he was only "going to get tougher" and advised them that he had many ways of keeping them in line.

"Don't forget the Army controls the food here," he told them.

Correspondents who crossed up the PRO and General MacArthur could also suddenly find that they were denied gasoline to run their vehicles. In addition, SCAP public relations officers would call editors back in the States to complain about reporters and demand substitutes. Killer Diller was eventually replaced by Brigadier General Frayne Baker, who was only marginally less hostile. General Baker told the Tokyo-based correspondents that they could be court-martialed for publishing classified material under the Articles of War because the United States and Japan technically were still at war. (At that moment, he informed them that SCAP officers could declare any information classified that it liked.)

Not even Hersey's war hero status would exempt him from close scrutiny—and SCAP was not the only U.S. government operation monitoring him. Tokyo-based Federal Bureau of Investigation officials were alerted to Hersey's arrival and relayed this information to the FBI director in Washington, D.C., with a request that the Tokyo office be furnished with further information about him. It is unclear if this was standard procedure, or whether Hersey's Moscow background had triggered the report request.

Further complicating Hersey's mission was the fact that even though some foreign journalists and photographers had been allowed to travel to Hiroshima and Nagasaki, the atomic cities were still a restricted topic. Under a press code issued by SCAP, the Japanese were hardly allowed to mention the bombings in poetry journals, much less mainstream publications, radio broadcasts, or scientific journals. Japanese reporters, photographers, and filmmakers who had been dispatched to the atomic cities in various capacities usually had their materials confiscated or hid them indefinitely for fear of having them taken or destroyed.

Hersey would have to tread lightly or risk being suppressed like his forerunner reporters Wilfred Burchett and George Weller—or worse.

YESTERDAY'S NEWS

Luckily for SCAP and U.S. government officials back in Washington, D.C., by the time Hersey came to Tokyo, few reporters were still vying to cover Hiroshima in any meaningful way. Likely most of them were simply too daunted to brave SCAP's restrictions, obstacles, and threats. Yet, just as members of the press had internalized racist propaganda against the Japanese during the war, many correspondents and editors in the months since the bombing also seem to have accepted in varying degrees the government line that the nuclear aftermath had been overly hyped.

Now that SCAP was clearing select journalists to visit Hiroshima and Nagasaki again, the stories they reported that spring depicted Japanese residents who appeared to have resumed near-normal lives. In February another official press tour had been taken to Hiroshima, this time to show journalists how quickly the city was recovering from the bombing. Lindesay Parrott, Tokyo bureau chief of the *New York Times,* reported in that newspaper that houses were being built and vegetable gardens being planted in the ruins. Visitors now had to "remind themselves as they surveyed the ruins that Hiroshima was the spot where atomic warfare stuck its first blow." Its residents had been surprisingly sullen, he added—except for those who ostentatiously paraded their bomb injuries for the correspondents.

"They seemed rather proud of such distinguished attention," Parrott stated, adding that all of the bomb-related injuries were now healing.

The message was clear: the emergency was over; there was nothing left to see here. Hiroshima and Nagasaki were yesterday's news. Joseph Julian, a radio reporter who had been to the city, told his boss that he wanted to do a series of broadcasts on Hiroshima and was discouraged.

"No one wants to hear about Hiroshima anymore," Julian was told. "It's old stuff."

Like the Bikini bomb test junketeers, the occupation press corps was becoming blasé and had long since turned their attention to new stories. Japan had become the land of limitless next scoops, and fewer journalists were interested in covering the aftermath of the last war. When Hersey arrived in Tokyo, many of his colleagues were busily covering the war crimes trials of General Hideki Tojo, former prime minister of wartime Japan, and his fellow defendants. The other big story being monitored: the threat of growing communism in Japan and the cultivation of the Japanese as a potential ally against the Soviet Union. Cold War rhetoric was heating up; former British prime minister Winston Churchill had just famously declared that the Soviets had erected an "iron curtain" across Europe. ("Nobody knows what Soviet Russia and its Communist international organization intends to do in the immediate future, or what are the limits, if any, to their expansive and proselytizing tendencies," he said, and cited the many countries over which Moscow was now laying claim.)

Japan was being given a rapid makeover from enemy state to eastern theater foothold for the United States against the USSR. ("Those Japs are going to be our allies in the next war," Hersey was told by an American intelligence officer. "I'll bet money on it.")

When Hersey applied to SCAP General Headquarters for clearance to travel to Hiroshima, there may have been some bemusement about why a reporter of his stature was asking to visit the site of the nearly year-old story. Yet the perception that the press corps was so subdued and distracted at this point—and that the story had been so successfully contained—probably worked to Hersey's advantage. Just two days after arriving in Tokyo, on May 24, he was informed by GHQ that he had been "authorized and invited to proceed by rail" to Hiroshima Prefecture.

Hersey immediately cabled Shawn, advising him that he was proceeding to Hiroshima. GHQ had given him a mere fourteen days on the ground. Once there, he would have to work fast.

Chapter Four

Six Survivors

KILLING FIELDS, PLAYING FIELDS

In 1946, the 420-mile train journey from Tokyo to Hiroshima Prefecture could still be a nearly twenty-four-hour ordeal in a packed railway car. One occupationaire commented that Tokyo's central train station was always filled with a "seething mass of people" and that "the air was heavy with thick, fleshy odor." The "coaches for Japs," he added scornfully, "with their close packing of humanity, have smells of all types that pervade the depot stations."

As for Hiroshima's train station, not much of it was left. Yet, because it stood slightly on the outskirts of the city, Hiroshima's decimated center could not immediately be seen by arriving travelers. Occupationaire rubberneckers who were taking the train through Hiroshima without disembarking sometimes complained about the "lack of A-bomb destruction," said one observer at the time.

Still, only the most cynical aftermath-tourist could be wholly dissatisfied by the scene at Hiroshima's station. Nearly a year after the bombing, the platform remained surrounded by splintered shells of former houses and buildings. A choppy sea of rubble—shattered wood planks, roof tiles, severed chunks of metal—spread out for miles around the station's platform. Blackened, shorn trees and charred telephone poles jutted from the earth.

When Hersey arrived in Hiroshima, temperatures were climbing; during the summer, the "ruins were terribly hot," recalled one blast survivor, who added that Hiroshima had few roofs or even walls to create shade. A "strange, unidentifiable odor" permeated the air, recalled another reporter who came to the city around the same time.

Hersey was stunned upon first beholding the city. He simply could not comprehend how so much damage had been done "by one instrument in one instant." The firebomb aftermath he'd just seen in Tokyo hadn't necessarily rattled him. "I'd seen damage like that in Europe and elsewhere," he said later. In any case, Tokyo's devastation had been achieved by waves of attacks by hundreds of planes. But Hiroshima terrified Hersey from the moment he arrived; the fact that a single bomb had caused this destruction would torment him throughout the duration of his assignment.

All along, Hersey had been aware of the irony that "a magazine of humor, and light pieces, and cartoons [was] suddenly devoting itself to something so terrible," as he put it. But now that he was on the ground, staring down the staggering devastation, post-bomb Hiroshima was no longer a horrible abstraction, a covered-up story to be broken in the pages of the *New Yorker*. Here, spread out in front of Hersey, were miles of jagged misery and three-dimensional evidence that humans—after centuries of contriving increasingly efficient ways to exterminate masses of other humans—had finally invented the means with which to decimate their entire civilization. Hersey worried that he wouldn't be able to bear the assignment after all, and resolved to do it as fast as he could.

Hiroshima had, over the last ten months, been attempting to rebuild, but reconstruction materials were scarce. Many survivors had tried to erect huts upon the ruins of their former homes. There were "teeming jungles of dwelling places, incredibly ugly . . . in a welter of ashes and rubble," recalled one reporter. Most of the makeshift homes were tin shanties, cobbled together from materials salvaged from the ruins. Makeshift signs had been nailed to the remnants of collapsed and burned former homes, detailing the whereabouts of its evacuated survivors or the fates of their now-dead previous occupants.

As residents attempted to clear lots to build new homes, they continued to uncover bodies and severed limbs. Around the time Hersey was there, in June, a concerted cleanup campaign in one district alone had unearthed 1,000 bodies.

The city's surviving population was starving. The American occupiers had brought into Japan shipments of corn, flour, powdered milk, and chocolate, but "supplies from the outside [to Hiroshima] were irregular," recalled another reporter in a diary he kept at the time. "The people here feel forgotten," he added. After Little Boy had destroyed the city, a typhoon and floods swept through the area and ruined the crops that had withstood the bombing. Some residents tried planting small gardens in the ruins next to their shanties. Vegetation was growing again in the atomic city: weeds, grass, and flowers were pushing almost aggressively through the rubble. The bomb had not only left certain plant roots alive underground, it had apparently "stimulated them," Hersey observed, with apparent revulsion. He noted that the aptly named panic grass and feverfew were among the plants finding the landscape most hospitable.

Getting around the city remained a challenge. At 8:14 a.m. on August 6, 1945, the city center streetcars had been filled with morning commuters. By 8:16 a.m. the streetcars in the city's center were distorted wreckages, filled with scores of blackened corpses. Some trams had been rehabilitated and were now chugging again through the partially cleared streets. A few residents pedaled on bicycles with tires improvised from all sorts of salvaged materials. Horse-drawn hearses still toted bodies away from the Hiroshima Red Cross hospital—which had, Hersey noted, managed to put up a new brick façade, even though the conditions inside remained remedial at best. (One visiting American doctor called the hospital conditions "absolutely horrible," adding that he would "sooner die than rot in one of [its] rooms.") Despite having been declared safe by General Groves and his team, several square miles were reportedly still roped off due to possible radioactivity.

The remains of a steel tower still stood at Hiroshima's ground zero. A sign affixed to it declared:

Center of Impact

The site was a shrine of sorts for the Japanese, but countless occupationaires had been making their way to ground zero for photo ops. A Planning Conference had been considering the creation of a monument to "International Amity," but for the moment, ground zero was being treated more or less as a theme park by visiting troops. Some grew almost giddy over all of the "bomb souvenirs" on offer in the ruins; not even the fear of possible residual radioactivity kept collectors away. "[The] treasure area involved hundreds of acres [containing] plenty of curios and family treasures mixed with the rubble," recalled one visitor, who scored several broken porcelain cups that he intended to use as ashtrays. (Others had much bigger hauls, he noted, but he was only an "amateur scrounger.") Despite having been fairly picked over, he added, there were still plenty of objects to choose from. He was certain that "a small fortune could be made" by selling them in America.

Occupation forces found other ways to turn the killing fields into playing fields—literally, in one case. Six months earlier, Marines stationed in Nagasaki cleared a space in that city's bomb debris to create a football field, where they played a New Year's Day "Atom Bowl," complete with a Marine band and goal posts made from salvaged scrap wood. They had also conscripted Japanese girls to act as cheerleaders. "We thought it would be totally appropriate," recalled one player years later, and that they'd felt "it would be great for publicity."

WHAT IT'S REALLY LIKE IN HEAVEN

Hiroshima offered few housing options, but Hersey found shelter at an American military police boardinghouse in Ujina, near the Hiroshima port. Situated about three miles away from the bomb's hypocenter, the municipality of Ujina had sustained less damage than areas closer to ground zero, allowing occupation forces to use intact buildings there. During World War II and previous conflicts, Japanese soldiers had been dispatched from the port to military theaters throughout Asia.

It would not be the first time Hersey lived with the U.S. military while researching a potentially damaging story. He had spent much of the war as an accredited war correspondent attached to various military units; those arrangements had not precluded him from scrutinizing the missions and personnel he was covering—nor from subsequently turning out critical assessments of them, as demonstrated in his withering General Patton portrait in *A Bell for Adano*. In any case, it was in Hersey's best interest to keep collegial relations with the Hiroshima-based U.S. officials, who would prove crucial in terms of procuring necessities, such as a jeep and gasoline. Hersey had been instructed in his invitational travel order that he would have to secure his own "subsistence," and if he planned to eat while in Hiroshima, he was more likely to get fed by the Americans than by the Japanese.

Hersey immediately began to scope out his protagonists, starting with the Jesuit priests he had read about in the Father Siemes testimony and heard about from Lieutenant Colonel McGovern and Lieutenant Sussan. Locating Hiroshima's resident Catholics was not difficult: the priests had been among the only Hiroshima residents who could afford building materials and had already begun reconstructing their small campus. First, they had cleared the ruins of their old mission compound and had erected barracks on the site. The logs piled up on the church's land plot for the planned mission buildings inspired envy in Hiroshima's less fortunate residents.

Hersey was already familiar with the head of the mission—Father Superior Hugo Lassalle, a German—whose ordeal had been recounted in the Father Siemes testimony reprinted in *Time* magazine's Pacific pony edition. Father Lassalle had been at the Society of Jesus's Central Mission and Parish House compound in Hiroshima's Noboricho neighborhood when the city was bombed. He currently resided, along with the other priests who had returned to rebuild their church, in a small one-room barracks structure made from galvanized iron sheets and wooden boards. The space was currently serving as a temporary church, reception room, and the priests' home. At night they slept on straw tatami mats on the floor; a piece of plywood at the front of the house served as a makeshift altar. Father Lassalle and his colleagues were subsisting on rice and Japanese white radishes.

Father Lassalle's floppy pale hair and oversized ears gave him an amiable appearance. Even among the ruins, he exuded determination and optimism. Having studied philosophy and theology in the United Kingdom before coming to Japan in 1929, he spoke English. When Hersey found and approached him about speaking with him for his article, Father Lassalle agreed.

On August 6, Father Lassalle recounted, he had been standing at the window of his second-floor room when the bomb's flash seared across the city. He managed to dart away from the window, which suddenly blew in. Shattered glass fragments speared his entire back and tore into his left leg; blood poured from the wounds.

"This is the end of me," Father Lassalle recalled thinking. "Now I will know what it's really like in heaven."

It was not, however, the end of Father Lassalle. Miraculously, even though the compound was a mere 1,400 yards from ground zero, the mission house's framework held fast; the structure had been reinforced by an earthquake-fearing priest in the order named Brother Gropper. No one else on the premises had died, either, although most had been pierced by broken glass and battered by splintered, flying furniture. Rather than discovering what heaven was really like, Father Lassalle quickly discovered that he and his colleagues had been ushered straight into hell on earth.

All of the other compound's buildings had collapsed, including a Catholic kindergarten near the front of the property. Children screamed from under the ruins. Father Lassalle and the other priests—including one who had received a serious head wound that was spurting blood—rushed to dig them out. They were finally "freed with the greatest effort."

In the immediate neighborhood, shocked, blood-covered survivors began to make their way into the street, which was littered with the debris of collapsed houses, fallen live wires, scalded telephone poles, and bodies. The priests tried to administer aid to as many people as possible until huge walls of fire started to rage toward the mission. As the priests fled the inferno, they heard neighbors and parishioners—trapped in the rubble of their collapsed homes—screaming in agony and calling for help. There was no way to rescue them; they had to be "left to their fate." They would all burn to death as the fires consumed the area.

DESCENT INTO HELL

Father Lassalle introduced Hersey to another German priest who had been with him that day: Father Wilhelm Kleinsorge, thirty-nine, who also spoke English and agreed to be interviewed. Father Kleinsorge had been very ill since the bombing. He had sustained only minimal injuries when Little Boy exploded, yet somehow Father Kleinsorge's small wounds had refused to heal over time, and he had been plagued by fever, nausea, diarrhea, and plummeting white blood cell counts. Around two weeks after the bombing, "I became extremely tired, and eventually I couldn't stand up," he later recalled. He made his way to a hospital in Tokyo, where "they told me it was very bad," and that his "bone marrow [had been] damaged by the radiation from the bomb." He returned to Hiroshima, where he had since been regularly in and out of the hospital.

Father Kleinsorge had blacked out when Little Boy exploded, but he remembered what happened beforehand and afterward in excruciating detail. When the bomb went off, he told Hersey, he was in a third-floor room at the mission house, reading a magazine, *Stimmen der Zeit* (*Voices of the Times*), in his underwear. Suddenly he saw a searing flash. At that point he blacked out. When he came to, he found himself staggering around outside in the mission's garden. Black smoke blotted out the sky above; all the buildings around him had collapsed. The other priests emerged from the house. Father Hubert Schiffer, the priest with the head wound, was bleeding so profusely that the others worried he would die.

Quickly the fires drew closer. If the priests lingered much longer at the house, "the oncoming flames [would have left] almost no way open" for them to escape. They were about to leave when they saw the diocese secretary, Mr. Fukai, standing and sobbing in his window on the house's second floor.

Hersey knew the broad strokes of the story of Mr. Fukai's grim fate from Father Siemes's testimony as well. Father Kleinsorge now filled in the details. He told Hersey how he had personally run back inside the house to retrieve Mr. Fukai. In hysterics, Mr. Fukai refused to leave the building, telling Father Kleinsorge that he could not bear to survive the destruction of his fatherland. Father Kleinsorge had to drag Mr. Fukai out of the house on his back and forcibly carry him away. Yet as the group of priests made their way through the streets, trying to navigate through the walls of flames and crushed buildings, Mr. Fukai managed to break away from them and run back toward the flames. He had not been seen or heard from since.

Father Kleinsorge and the other exhausted priests took refuge in

Asano Park, a riverside estate. Hours earlier, the park had been an elaborate Japanese garden; it was now a backdrop for unspeakable horrors. Hundreds of other survivors were also making made their way there; the dead and the dying covered the grounds. Some had been so badly burned that their faces appeared blurred, like smeared paintings. Bomb-induced winds were tearing across the city, and a violent whirlwind emerged near the park; it began tearing up surviving trees in its path and twirling them in the air. It then moved into the river, forming a water spout 100 meters high. As the storm moved away, it swept terrified refugees into the water.

A group of priests from Father Siemes's novitiate, a little more than a mile outside the town, had heard that the city center priests were taking refuge in Asano Park and had come to find them, makeshift stretchers in hand. Father Siemes's team were appalled by the scene that awaited them in the city.

"As far as the eye could reach, [Hiroshima] is a waste of ashes and ruin," Father Siemes said. "The banks of the river [were] covered with dead and wounded, and the rising waters [of the city's rivers] have here and there covered some of the corpses." Survivors dragged themselves under burned-out cars and trams, seeking shelter from the fires. "Frightfully injured forms beckon[ed] to us and then collapsed. Hiroshima's broad main street was littered with naked, charred cadavers.

Even more horrific experiences were in store for Father Kleinsorge, who remained in Asano Park after the other priests had been carried out to safety. As he left the gardens to search for fresh water, he had to step over dozens of festering, blistered, and peeling human bodies. He found a working tap nearby. While ferrying water back to the park, he encountered a large group of Japanese soldiers, all desperate for a drink. Their eyes had melted away in their sockets; the liquid had run in rivulets down their faces, which were burned beyond recognition.

Father Kleinsorge eventually made his way to Father Siemes's novitiate with the other priests. The mission took in fifty refugees but had few supplies to tend their wounds and feed them. As the days passed, Father Kleinsorge started to develop soaring fevers and was sent to Tokyo's International Catholic Hospital. His doctor told the priest that he could expect to go home in a couple of weeks. But then, when the doctor went into the hallway, Father Kleinsorge overheard him say to someone else that he expected the priest to die.

"All these bomb people die," he told the hospital's Mother Superior. "They go along for a couple of weeks and then they die."

RESCUING ANGEL

As the two men spoke, Hersey clearly earned Father Kleinsorge's trust and faith, for the priest not only agreed to introduce Hersey to other blast survivors—or *hibakusha* ("atomic bomb–affected people")—but also offered to act as a translator for him.

The Father Siemes testimony that Hersey had read also mentioned an unnamed Japanese Protestant minister, a "rescuing angel" who had helped evacuate the wounded Catholic priests from Asano Park. This gentleman, Father Kleinsorge told Hersey, was Reverend Kiyoshi Tanimoto, Methodist pastor from Hiroshima.

Fortune continued to favor Hersey, for Reverend Tanimoto, thirty-six, also spoke English, having studied in the United States: he had graduated from the Candler School of Theology at Emory University in Atlanta, Georgia, a year before the Japanese attack on Pearl Harbor. Father Kleinsorge offered to bring Hersey to meet him.

Like the Jesuit priests, Reverend Tanimoto had returned to live among the ruins and was determined to rebuild his own church—which had, along with his home, collapsed and burned on the day of the bombing. He was currently renting a leaking house in the Ushita section of Hiroshima—a less-damaged suburb in the northern part of the city—for himself; his wife, Chisa; and their infant daughter, Koko. On the house's front wall he had nailed a handmade sign:

Temporary hall and parsonage, Hiroshima Nazarekawa,
Japan Church of Christ

Chisa Tanimoto greeted Hersey and Father Kleinsorge. Reverend Tanimoto was not at home, she told them. He was out trying to secure materials for reconstructing his church and was conducting services for surviving residents around the city. Hersey left his business card.

Reverend Tanimoto was exhausted when he returned home that night. Like Father Kleinsorge, he had been chronically ill since August. The high fevers, bed-soaking night sweats, diarrhea, and total debilitation had begun a few weeks after the explosion of the bomb.

"I [have] contracted the atom disease," he had recorded in his diary that previous fall.

His rented house stood near the office of a doctor, who administered vitamin shots to the pastor but could do little to alleviate his suffering. Reverend Tanimoto's family feared that he would die. Scores of other blast survivors around them had died in agony after losing their hair, vomiting blood, and developing frightening red-black blotches on their skin. Fortunately, Reverend Tanimoto had managed to persevere. Unfortunately, his painful symptoms did as well.

That evening Chisa handed him Hersey's business card. "Father Kleinsorge of the Catholic Church has brought [the] gentleman of this card," she told her husband. Reverend Tanimoto examined it. Next to his name, Hersey had scribbled "*Life*" and "*The New Yorker*." Reverend Tanimoto was familiar with both publications. Chisa explained that Hersey had hoped to interview the pastor about his experience on the day of the bombing, and that the journalist would return the following day.

"Entirely ignorant of his being a prominent rising American writer, I felt no particular interest in his visit," Reverend Tanimoto recalled later. "Up to that time, I had actually met a few American reporters and then offered them the material of my experience in the atomic bomb detonation. I had never heard from them thereafter and, therefore, I was rather losing interest in meeting a news correspondent."

Still, something about Hersey's card made Reverend Tanimoto think again. He respected both *Life* and the *New Yorker*. Perhaps this reporter would be different. And after all, Hersey had gone through all the trouble to seek him out. Responding to his kindness was the polite thing to do, Reverend Tanimoto reasoned.

The pastor had to go out again the following day, so he decided to write Hersey a letter instead, giving an overview of his ordeal the previous August. Despite his exhaustion, Reverend Tanimoto sat up in his bed until 3:00 a.m. writing Hersey a ten-page missive in English, detailing the horrific scenes he had witnessed the day the bomb had been dropped. If Hersey was interested, Reverend Tanimoto ventured, he could personally take the reporter to the places he described. With the letter, which he brought to Father Kleinsorge early the next morning, he included a hand-drawn map of the city indicating some of these locations.

Father Kleinsorge arranged for the two men to meet at the Jesuit priests' temporary headquarters. That Saturday morning, June 1 at 9:00 a.m., Hersey arrived to interview Reverend Tanimoto. Hersey was wearing his military-issue war correspondent's uniform, but "unlike a soldier," thought Reverend Tanimoto, "[Hersey] had about him the refinement of a literary man."

"I've read your [letter]," Reverend Tanimoto recalled Hersey saying, "and I think it's quite moving." Hersey explained that he wanted to tell the story of the bombing of Hiroshima not from a scientific point of view but rather "from the standpoint of humanity."

Reverend Tanimoto began to talk. "[Hersey] listened to me intently," he observed, "and showed such sympathies that I spoke to him quite frankly."

THE CHARON OF HIROSHIMA

The pastor had gotten up at 5:00 a.m. on August 6, 1945. Exhausted from another night of air raid warnings, he made himself a breakfast of soybean powder and rice bran mixed with water; substantial food had long been scarce in the city. There had been air raid alarms in Hiroshima nearly every night for weeks. The city's residents were confused about why they had thus far been spared the heavy bombing that had decimated Japan's other major industrial cities, but lately they'd started hearing "fantastic rumors that the enemy had something special in mind for [the] city," as Father Siemes put it. The city had been lined with fire prevention lanes in case American B-29s made firebombing raids like the ones that had been pulverizing Tokyo.

Fearful of possible raids, Chisa and baby Koko had been spending nights at a friend's house in a neighborhood removed from the city center. Reverend Tanimoto had started moving treasured objects from his church to the summer house of a wealthy manufacturer two miles away for safekeeping. He had already carted a stash of clocks, Bibles, church records, and altar objects to the residence, and even the church piano and organ. In his pushcart on the morning of August 6 was a heavy cabinet. Together with a friend, he'd hauled the piece of furniture two miles away from the city center and up a hill toward the safe house.

At 8:15 a.m., as Reverend Tanimoto stood in front of the porch of the summer house, he saw a "sharp flash of light streak out of a dim, brown cloud . . . [A] strong blast of wind filled the air." In terror, he fell to the ground between two rocks as splintered wood and glass rained down on him. An earthen storehouse had stood between him and the blast. When he raised his head, he saw that the summer house had collapsed and the houses around him were on fire.

Neither he nor his friend had been injured. Reverend Tanimoto climbed a nearby hill and saw that the city below was engulfed in flames. Black clouds swirled violently above the city center. A line of stunned, blood-covered survivors began staggering out of the city, heading up the hillside road.

"Most of the people were naked," Reverend Tanimoto recalled. "Skin from faces and hands, arms and breasts was stripping off or hanging loose . . . It seemed like a procession of ghosts."

Terrified about the fate of his little family and his parishioners, Reverend Tanimoto ran toward the city. All of the houses along the way had been heavily damaged but were still partly standing. However, as he got within one kilometer of the city center, the buildings "were all flat on the ground as if they had been merely toy buildings crushed by a hammer." He heard "painful cries of *Tasukete!* Help!' from under smashed houses." He ran toward his own now-demolished neighborhood, Noboricho, but furious fires blocked most of the ways in. Hundreds of bodies and dying people lay in the streets, all about to be consumed by the flames. Reverend Tanimoto came across a stash of cushions and, after soaking them in a water tank, covered himself with them and plunged through the fires, growing totally disoriented amidst the inferno. He finally managed to get into his neighborhood, only to be greeted by a sudden whirlwind.

"Red hot iron sheets and burning boards were spiraling in the air," he later recounted. The whirlwind lifted the pastor some seven or eight feet off the ground as though he were swimming through the air. It abruptly dropped him on the ground, knocking the breath out of him. Suddenly he heard enormous explosions nearby: gasoline tanks were exploding.

Astonishingly, Reverend Tanimoto found Chisa in a line of refugees staggering away from the city center. In her arms, she was carrying Koko, who had also survived: "[Chisa] was in a blood-stained chemise, . . . her hair was hanging about her shoulders, and she had a ghostlike appearance." She had returned with the baby to the parsonage that morning; when the bomb went off, she was holding Koko and standing in the front doorway, talking with another church member. The building immediately collapsed on top of them, burying them under a pile of heavy wood and rubble.

Crushed under the weight of the wood, Chisa had lain unconscious with Koko pressed to her chest. The baby's cries finally brought her back to consciousness, and "she struggled with all her strength in order to save her baby," Reverend Tanimoto said. Her arms were pinned to her sides, but she managed to free one and scratch out a hole in the debris. Soon it was just big enough to push Koko through, and eventually, miraculously, Chisa managed to drag herself out as well. They emerged just in time: fire was closing in. They fled in the direction of Asano Park, where the Catholic priests had also evacuated, and encountered Reverend Tanimoto on their way there.

The pastor was shocked by the scene in the park. Among the refugees he found there was his next-door neighbor, a young woman named Mrs. Kamai. Reverend Tanimoto's own baby had survived; Mrs. Kamai, however, was clinging to her dead infant daughter, who had choked to death on dirt when they were trapped in the debris of their collapsed home. She would not surrender the child's body for days, even after it began to decompose in the summer heat.

The pastor decided to try to bring more refugees across the river to Asano Park. He saw a small boat beached on the rocks, only to discover that it was filled with five corpses whom he had to drag out, apologizing to each man and praying as he did so. Pushing the boat with a pole, he began to transport gravely wounded survivors to the opposite bank like Charon ferrying the souls of the deceased across the river Styx. As he tried to pull one man into the boat, he saw with horror that "the skin of his hand slipped off as if it were a glove."

"I had no more human feelings," he stated. "Slowly pushing aside the dead [bodies in the river], I propelled my boat."

In the park, Reverend Tanimoto encountered Father Kleinsorge and the other Catholic priests, his neighbors in Noboricho. They had taken refuge under some bushes, except for Father Kleinsorge, who was distributing water to survivors. When Father Siemes and his priests arrived, Reverend Tanimoto helped them transport the wounded priests by boat upstream to a point where their colleagues could carry them to safety on litters. After nightfall it had become difficult in the darkness to keep from stepping and slipping on the bodies covering the ground. The river tide rose and swept away more corpses—and drowned others who were still alive but too weak to move away from the water.

Around midnight Reverend Tanimoto lay down briefly in the park. A severely burned young girl lay nearby; she was a student who had been requisitioned that day, along with her schoolmates, to help tear down houses to make way for fire lanes. She shivered in misery.

"Mother, I'm cold," she whimpered. "Mother, I'm cold."

No one that day had any inkling of what had hit them, the pastor noted, or even knew what an atomic bomb was—with a few exceptions. The deputy director of the Hiroshima's Red Cross Hospital, Dr. Fumio Shigeto, had taken a picture of the destruction and run down into the hospital's dark room to develop his film. To his surprise, he found that it had already been exposed—a first clue that this weapon had had unusual qualities.

After Hersey and Reverend Tanimoto had talked for about three hours, they walked out of the house and into the neighborhood. The two men went to the site of Reverend Tanimoto's former church, still in ruins. He was having far less luck than the Catholics in securing rebuilding resources. Hersey had brought along a camera, and he asked Reverend

Tanimoto if he could take pictures. Hopefully he would have more luck getting his film out of Japan than Wilfred Burchett, whose Hiroshima film had gone missing under suspicious circumstances after his "Atomic Plague" story was released in the *Daily Express*.

Like Father Kleinsorge, Reverend Tanimoto agreed to make further introductions for Hersey among Hiroshima's blast survivors. Even though Hersey had been a stranger to Reverend Tanimoto just hours earlier—a reporter from the nation that had just destroyed the pastor's city, and defeated and occupied his country—Tanimoto had the peculiar feeling that he had been speaking with "an acquaintance I had missed for years."

TEN THOUSAND PATIENTS

In the days that followed, Father Kleinsorge and Reverend Tanimoto approached survivors they knew, asking them if they would be willing to speak with Hersey. Day after day Hersey listened to their accounts, internalizing all that they had been through and seen, taking notes throughout. He may have used shorthand, which he had learned when briefly serving as an assistant to writer Sinclair Lewis. Others recalled Hersey simply listening to their testimonies and taking the information in.

Hersey seemed to lose track of the number of people he ultimately spoke with. His later estimates would range between twenty-five and fifty survivor interviews. Each story contained its unique horrors. He held fast to the example of *The Bridge of San Luis Rey* as a way of choosing his subjects: all of them had to have overlapped on the day that Little Boy had torn apart their city and lives. Gradually, the short list suggested itself. In addition to Father Kleinsorge and Reverend Tanimoto, Hersey ultimately chose two Japanese doctors, a young female Japanese clerk, and a Japanese widow, mother to three young children.

Mrs. Hatsuyo Nakamura had been a neighbor of the Jesuit priests and Reverend Tanimoto in Noboricho. Her husband, a tailor, had been drafted into the Japanese army and died in Singapore in 1942. Since then Mrs. Nakamura had been supporting their three young children by taking in sewing jobs, which she completed on her dead husband's sewing machine. The children—one son and two daughters—were ten, eight, and five years old the day the bomb fell. The youngest—a little girl named Myeko—had attended the Catholic kindergarten at the Society of Jesus mission.

When Hersey was introduced to her, Mrs. Nakamura was also living once again in Noboricho. She had been poor for years, but had, since the bombing, learned the true meaning of destitution. She was dwelling in a one-room shack with a dirt floor; holes in the walls had been stuffed with paper and cardboard. Meager meals were eaten from plates and utensils salvaged from the same ruins that were being gleefully pillaged for souvenirs by occupationaires. After clearing a patch of rubble, she had scratched a vegetable garden into the earth near the shack. Like Father Kleinsorge and Reverend Tanimoto, Mrs. Nakamura had been badly afflicted by what Reverend Tanimoto had called the "atom disease": all of her hair had fallen out after the bomb, and she had nearly wasted away from post-bomb diarrhea and vomiting.

Mrs. Nakamura's shack was so tiny that it could barely contain Hersey when he came to interview her. "He sat on the floor . . . with his feet propped up in front of him, and it seemed as though his legs filled the entire room," she later related.

He was gentle and friendly, and put her at ease as she recounted the day of the bombing. The moment that the bomb exploded, she told him, she was in her kitchen cooking rice and looking out her window. Her three children were still sleeping on bedrolls in the next room, exhausted after a false-alarm air raid evacuation in the middle of the night. When the bomb went off, she was catapulted into the next room as tiles and timbers fell on her from above.

Her three children were buried in debris. Mrs. Nakamura frantically dug them out. They were stunned but not injured. When the Nakamuras managed to make their way into the street, they were bewildered and terrified by the suddenly black sky, the smashed houses, and soaring fires around them. They ran through the chaos to Asano Park. On the way they saw Myeko's smashed kindergarten and also glimpsed Father Kleinsorge, stunned and bloodied outside the Catholic mission.

Once at the park, the entire family began throwing up. They drank river water and threw up again, and continued to vomit uncontrollably. The Nakamuras had been among the first arrivals. The death scene there quickly unfolded before them as disfigured blast survivors filed into the park and died by the scores in front of them. Then the terrifying whirlwind tore through the area.

At one point a Japanese naval launch passed by, from which an announcement was broadcast: a naval hospital ship would be coming soon. Mrs. Nakamura was briefly reassured: a doctor would be there before long to treat her children. But this naval ship never materialized, and most of Hiroshima's doctors and nurses were dead or gravely wounded: 270 of the city's 300 doctors had died in the blast or were injured; of 1,780 nurses, 1,654 had been killed or injured. Most of the surviving doctors had suffered serious wounds: one physician, from the partially burned Hiroshima Communications Hospital, tended countless patients despite having been slashed by more than 150 shards at the moment of detonation.

The surviving medics were confused by the prevalence of diarrhea and vomiting among the thousands of people converging upon the few medical facilities that were still standing. Had the bomb released a poison gas or perhaps some deadly microorganisms? wondered one, who erroneously concluded at first that he was seeing scores of cases of bacillary dysentery. In any case, the hopelessly outnumbered medical workers were soon unable to treat any condition: their supplies of bandages, drugs, and clean water were quickly spent.

Through Father Kleinsorge and Reverend Tanimoto, Hersey was introduced to the only medic from Hiroshima's Red Cross Hospital who had survived the blast unscathed. Dr. Terufumi Sasaki, a young member of the surgical staff, had been on-site when the bomb went off and witnessed the bomb's physical effects in the immediate aftermath of the explosion. His medical observations about the bomb's effects would be crucial to Hersey, and it was imperative that Hersey get all the medical terminology exactly right in his report. To that end, the reporter questioned the multilingual Dr. Sasaki with the help of three Japanese interpreters; they also spoke in German, with the help of Father Kleinsorge; as well as in Mandarin, for Dr. Sasaki had trained in China; and in English.

Dr. Sasaki, who was in his mid-twenties, told Hersey that he had been walking down the hospital's main corridor when he saw the flash. Ceilings suddenly collapsed; beds were sent flying. Blood was splattered across the walls. Shattered glass and the bodies of dead patients and medics covered the floors. Dr. Sasaki's glasses were blown off his face; he took a pair off the face of an injured nurse. He immediately gathered all of the unbroken and unburied supplies he could find and began bandaging the wounded hospital staffers and patients.

Blast survivors began to stream in from outside. Soon they crowded the rooms, hallways, stairwells, front steps, bathrooms, and stood or lay by the hundreds around the building. Ten thousand casualties had eventually converged upon the Red Cross Hospital, with its six hundred already full beds and one uninjured doctor. A doctor at the Hiroshima Communications Hospital similarly recalled that people "came as an avalanche and overran the hospital" and that they were "packed, like rice in sushi, into every nook and cranny . . . [Cleaning] the rooms and corridors of urine, feces, and vomitus [became] impossible." Outside, the front steps to that hospital became slick with excrement. Soon there would be the smell of decomposing dead bodies as well, for there was no one on hand to clear them. The hospitals quickly became ringed by hundreds of corpses.

Dr. Sasaki worked for nearly seventy-two hours straight. Eventually another doctor and twelve nurses arrived from another city, but the medical team was still overwhelmed by the task of caring for so many victims. By the end of the third day, many of the patients Dr. Sasaki had attended to earlier were dead.

THE DOCTOR WHO DIDN'T COME

Not all of the survivors whom Hersey would profile acted as selflessly on the day of the bombing. He was also introduced to another Hiroshima doctor, one who had felt too injured in the bombing to help the dying masses. Dr. Masakazu Fujii, the proprietor of a small, single-doctor private hospital in Hiroshima, had been a neighbor of Father Kleinsorge and the other Catholic priests. Only a few days before the bombing, he had given them a first aid kit.

After the bombing, Dr. Fujii opened a new private practice in a Hiroshima suburb. By the time Hersey met him, the doctor had hung a sign outside his new practice, announcing:

M. Fujii, M.D.
Medical & Venereal

Hersey noted that the sign had been written in English, as Dr. Fujii had become solicitous of the American occupationaires. His small private hospital had been destroyed in the bombing, but in the bomb's aftermath, Dr. Fujii had more than his share of desperately ill patients and was able to reestablish himself quickly.

When Hersey came to speak with Dr. Fujii, the doctor did not understand at first that Hersey was a reporter. Many Americans had come to Hiroshima to interview blast survivors in recent months. "Some were doctors; others were investigators for the American or Japanese government," he later noted. "I thought Mr. Hersey was one of them." Hersey gave him his calling card, but the doctor did not recognize that the publications listed on them were American magazines.

The doctor spoke with Hersey for three hours, with the help of Father Kleinsorge. Dr. Fujii immediately felt admiration for the reporter.

"When he was here speaking to me, and when I looked at him, I said to myself, 'This young man is the type of American the great President Lincoln must have been,'" the doctor recalled. "He even [looked] like the pictures of Lincoln I have seen in the history books . . . and he [was] so understanding and sympathetic."

On the morning of August 6, 1945, Dr. Fujii, like Father Klein-sorge, had been reading a newspaper in his underwear—in Dr. Fujii's case on the porch of his hospital, which had been built out over one of Hiroshima's seven rivers. When the bomb went off, he was plunged into the water, the wreckage of his hospital crushing down on top of him. Instead of drowning, however, Dr. Fujii was suspended between two timbers—"like a morsel suspended between two huge chopsticks," as Hersey would put it—with his head above the water. The prospect of being submerged when the tide came in seemed to imbue the doctor with sudden strength: despite his broken collarbone and likely fractured ribs, he was able to free himself. Two of the nurses from his hospital were also suspended in the timber like flies in a spider's web. With help from survivors on the bank, Dr. Fujii was able to free them. His other staff, patients, and a niece who was staying with him had perished in the hospital.

Nearby, as the injured Catholic priests emerged bloodied and bewildered from their mission, they tried to reach Dr. Fujii's hospital before fleeing to Asano Park. Dr. Fujii's hospital was only about six blocks away, but the soaring fires had engulfed the streets in-between. As the flames swept toward his demolished hospital building, Dr. Fujii and the nurses plunged back into the river. There they waited for the conflagration to die down, until the tides rose and the waves became choppy. Although injured, the small group managed to make its way upstream and took refuge on a bank near Asano Park.

Throughout that afternoon and night, amidst the Asano Park death scene, Reverend Tanimoto grew angry about the lack of emergency medical care for the wounded and dying. He left the park at one point and accosted a Japanese army medical unit working at another evacuation site, trying unsuccessfully to get an army doctor to come back with him to the park. Meanwhile, Dr. Fujii quietly left the bank where he'd been resting and ultimately made his way to a friend's summer house in a village outside of town, where he recuperated.

The Catholic priests at Asano Park were, at least, able to bandage themselves with the supplies that Dr. Fujii had given them earlier that week.

CRUSHED BY BOOKS IN THE ATOMIC AGE

Dr. Sasaki, Father Kleinsorge, and Reverend Tanimoto all confessed to Hersey that, after a certain point on the day of the bombing, after witnessing so much horror, they had simply gone numb, shut down emotionally. As Hersey heard harrowing account after account, he, too, may have become overwhelmed and hit his saturation point. One particular testimony, however, stood out for Hersey in its irony.

Before the bombing, Miss Toshiko Sasaki, twenty, had lived with her family in a Hiroshima suburb and commuted to her job as a clerk at the East Asia Tin Works in the city. When Hersey met Miss Sasaki, she was still recovering from bomb injuries at the Red Cross Hospital. After her ordeal—not only had she been hospitalized for months but her parents and infant brother had died in the bombing—she took counsel and received solace from Father Kleinsorge, who saw in her a candidate for conversion. Miss Sasaki had fallen into despair. On top of her other troubles, Miss Sasaki had also been rejected by her fiancé. He had been drafted to China during the war but had since returned alive. Yet now he apparently wanted to get out of the engagement. Many atomic bomb survivors had been deemed defective after the disaster and treated as pariahs.

Despite her physical pain and depression, Miss Sasaki also agreed to tell Hersey her story. The morning of August 6, she had just arrived at the tin works and sat down at her desk when the blinding flash came.

The ceiling caved in; several tall bookcases pitched forward and dumped all of their heavy books upon Miss Sasaki. One of the shelving units fell on top of the pile, further crushing her. She was trapped beneath them, her left leg twisted and badly broken. Unlike Mrs. Nakamura or Dr. Fujii, she could not free herself from the pile of volumes, plaster, and splintered wood, and languished there, in excruciating pain, drifting in and out of consciousness for hours.

A man eventually appeared and was able to pull some of her colleagues out of the rubble, but he couldn't get Miss Sasaki out and abandoned her there. She was later extracted by several other men, who left her sitting in the tin works courtyard in the rain. She was then deposited by another "helper" in a nearby makeshift shelter made from a sheet of corrugated iron against a courtyard wall. Her lower left leg had been pulverized and dangled from the knee. The pain was horrific. Soon two other disfigured survivors had also been wedged into the shelter, including a man whose face had been terribly burned. The "three grotesques," as Hersey would later describe them, were left there alone to fester and starve for forty-eight hours.

Eventually they were discovered. Miss Sasaki was transported by truck to an Imperial Japanese Army relief station. Her leg was by then enormously swollen and filled with pus, and although she was eventually shuffled among other army hospitals and temporary aid stations, no doctor could set the breaks—although one team of doctors eventually affixed it to splints.

Miss Sasaki eventually came under Dr. Terufumi Sasaki's care. By the time Hersey met her, she had healed considerably and could hobble on crutches, but her left leg—having never been set properly in those early weeks—was now three inches shorter than her right leg. She had been having arguments with Father Kleinsorge over the curious nature of his God. He was supposed to be merciful, she said. So why did He allow such extreme suffering? Father Kleinsorge's answers must have been convincing: that summer, in 1946, Miss Sasaki decided to convert to Catholicism.

LEAVING SILENTLY

Hersey had now identified his six protagonists. Through them he could tell the story of the bomb's true human cost at last. He acknowledged that his short list was "by no means representative of a cross section of Hiroshima's population," yet he felt that "the kinds of consequences of the bombing that they suffered were probably fairly true of the way the shadow fell on everyone."

After about two weeks on the ground, Hersey prepared to return to Tokyo, then to fly back to the United States to write his story. Some of the military police officers he stayed with knew that he was meeting and talking with bomb survivors, but they apparently made no attempt to interfere. Perhaps by then they were simply used to SCAP-approved reporters coming through the area to observe the devastation, speak with survivors, and then leave quietly. Just a few weeks after Hersey left Hiroshima, another small group of reporters came in jeeps to survey the damage now that the first anniversary of the bombing was imminent. The correspondents gawked at the destruction, as the previous reporters had, and even interviewed an unnamed doctor at Dr. Sasaki's Red Cross Hospital who'd been on-site the day of the bombing. But no ground-breaking or even faintly revelatory reporting resulted from this trip. Back in Tokyo, the occupation press corps was still closely monitoring the Japanese war crime trials that were in progress and busily debating whether the Japanese emperor should have been put on trial as well and "hanged by the neck until dead."

Hersey was still on track to "scoop the world," as Harold Ross had put it.

As Hersey stood on the platform of Hiroshima's train station, he thought about Miss Sasaki, pitifully broken in the ruins of the tin factory. How ironic, he thought, that she had been crushed by books in the first seconds of the atomic age. He would have to write a line to that effect in his story, he decided.

After Hersey left the city, Miss Sasaki and the five other survivors returned to their daily lives in their violently rearranged new world. Reverend Tanimoto began to prepare memorial services that would take place on the grim anniversary of the bombing. Dr. Fujii saw his private clients and entertained occupationaires, plying them with whiskey. Father Kleinsorge's health declined again, and he checked back into a hospital in Tokyo, this time for a month. Mrs. Nakamura's hair was growing back again; her five-year-old was back in school, for the Catholics had managed to reopen a kindergarten; her other children were back at primary school, which had to be held outside due to the lack of intact buildings in the neighborhood. Dr. Sasaki continued his work at the hospital.

The ruins of Hiroshima baked in the summer heat. None of the six interviewees had any idea that, in just a matter of weeks, their names and stories would become known around the world.

Chapter Five

Some Events at Hiroshima

A DELIBERATE SUPPRESSION OF HORROR

While Hersey was doing his final interviews in the bombed city, Harold Ross and William Shawn were half a world away both geographically and metaphorically. Their latest issue featured a profile on a hot dog–loving New York–based meat tycoon and a story on the Astoria Stakes, cushioned by cartoons and ads for Elizabeth Arden face cream, Coca-Cola, Underwood tinned deviled ham, and Lincoln Continental Cabriolets. Nothing in the magazine—now fully back in peacetime mode—foreshadowed the story that they would soon run.

On June 12, Shawn received a cable from Hersey, sent from Tokyo. The Hiroshima information had been successfully gathered, Hersey reported; he was leaving later that day and would be home in New York in about five days.

Hersey was given a seat on an interminable Air Training Command flight from Tokyo to the United States, with a stop in Guam and then another layover at Hickam Field, the principal U.S. Army airfield on Hawaii. Hickam stood adjacent to the U.S. Naval Station Pearl Harbor, on the island of Oahu, and had been attacked during the 1941 Japanese assault on Pearl Harbor. Hickam's hangars, barracks, and chapel, among other facilities, had been bombed. Like the Hiroshima bombing mission, Japan's Pearl Harbor attack had been a surprise morning offensive. Thirty-five men died instantly when a Japanese bomb hit the Hickam Field mess hall during breakfast. The layover at the base was a grimly appropriate bookend to Hersey's trip.

Finally, the ATC plane landed in San Francisco. From there Hersey made his way back to the East Coast. The deadline pressure was on: his story would run in the *New Yorker* to commemorate the August 6 anniversary of the Hiroshima bombing. He was used to writing articles and books in a "white heat," as he would later put it, but this would be a complicated, nerve-racking write, given the contentiousness of his material and Ross and Shawn's rigorous editing process. Plus, there was the possibility of competitive stories coming out in other publications around the anniversary.

Hersey began to write. At the top of a sheet of paper, in pencil, he jotted down a possible title for the story: "Some Experiences at Hiroshima." He crossed it out. He tried another: "Some Adventures at Hiroshima." That didn't work either. Nor did "The 'Original Child' Bomb"—a rough translation, Hersey was told, of the initial Japanese term (*genshi bakudan*) for the new weapon that had been unleashed against them. He settled on the provisional title "Some Events at Hiroshima."

Now that Hersey had his survivor testimonies in hand, he gamed out his approach to make his story as engrossing as possible. He felt that the article would have to read like a novel. "Journalism allows its readers to witness history," he later said. "Fiction gives readers an opportunity to live it."

His goal with the Hiroshima story was "to have the reader enter into the characters, become the characters, and suffer with them." Readers would have many excuses to put the magazine down quickly: the material might be too graphic and disturbing; they might have fraught consciences about the bombing itself; the story might seem preachy. Overcoming willful avoidance was one of Hersey's several high-stakes challenges. He needed to create something that readers simply could not put down. If his article read like an engrossing albeit horrifying thriller, he might have a shot at commanding the public's attention.

In the spirit of *The Bridge of San Luis Rey*, Hersey planned to interweave the day-of-bombing narratives of Father Kleinsorge, Mrs. Nakamura, Reverend Tanimoto, Dr. Sasaki, Dr. Fujii, and Miss Sasaki, using cliff-hangers throughout the article. He also decided to present the survivors' stories in a subdued but unflinching manner. In the past, Hersey's journalism for Time Inc. could be stylized and laced with pronouncements, with a tendency toward omniscience. All of that would be left behind with this article.

"My choice was to be deliberately quiet in the piece," Hersey later stated. This intentional "suppression of horror," he added, would create "an effect far more morally disturbing [than] would have been achieved had I shouted or screamed my outrage."

The writing would be stripped down, an almost clinical presentation of the action and facts—the opposite stylistic approach from the *New York Times*' "Atomic Bill" Laurence, who was at that moment preparing to release his own (military-approved) book, *Dawn Over Zero*, about the evolution of the nuclear bomb and his experience as General Groves's anointed in-house historian. At the moment the first bomb was detonated in New Mexico, Laurence wrote, "the hills said yes and the mountains chimed in yes. It was as if the earth had spoken and the suddenly iridescent clouds and sky had joined in one affirmative answer. Atomic energy—yes. It was like the grand finale of a mighty symphony of the elements, fascinating and terrifying, uplifting and crushing, ominous, devastating, full of great promise and great fore-bodings."

Enough of that. Pencil in hand, Hersey sat down and wrote—in tidy, calm penmanship—a draft of what would become one of the most famous, simple ledes in journalism history:

> *At exactly fifteen minutes past eight in the morning, on August 6, 1945, Japanese time, at the moment when the atomic bomb flashed above Hiroshima, Miss Toshiko Sasaki, a clerk in the personnel department of the East Asia Tin Works, had just sat down at her place in the plant office and was turning her head to speak to the girl at the next desk.*

He detailed where each of his six interviewees had been at the exact moment of detonation, and added that they did not know why they had been spared when thousands of others had died. "And now," he wrote, "each knows that in the act of survival he lived a dozen lives and saw more death than he ever thought he would see."

In granular detail, Hersey then lay out the moments in each survivor's morning routine, leading up to the bombing, and what happened to each immediately after the bomb exploded. (Hersey initially titled his first chapter "The Flash" and then later changed it to "A Noiseless Flash," remembering that none of his protagonists had actually recalled hearing an explosion; the blinding flash was the detonation detail that each remembered.)

He told of Reverend Tanimoto beholding the ghoulish procession of bloodied, shocked blast survivors streaming out of the city; of Father Kleinsorge and Dr. Fujii reading newspapers in their underwear at their respective homes when Little Boy rained death upon the city; of Dr. Sasaki's arrival at the hospital that morning (he had come in earlier than usual; had he taken his usual train and streetcar, he likely would have been standing near ground zero when the bomb went off), and of the horrific death scene at the Red Cross Hospital.

Hersey wrote of Mrs. Nakamura and Miss Sasaki tending to their families in the early hours of the morning, and of Mrs. Nakamura's desperate search for her children in the ruins of her collapsed house. He wrote of Miss Sasaki's ordeal at the tin works when the ceiling and bookshelves fell upon her. Hersey ended this initial section with the line he had thought of while standing on the Hiroshima train station platform: "There in the tin factory was a situation for the first moment of the atomic age: a person crushed by books."

He began the second chapter, which he titled "The Fire," and here the reader would truly become immersed in the horror wrought by the bomb. In this section, in nearly deadpan language, Hersey documented Reverend Tanimoto's descent into the inferno to try to find his family and neighbors; the heartbreaking fate of Mr. Fukai, the mission secretary who had committed suicide by running back into the flames engulfing the city; the Nakamura family's flight to Asano Park. He wrote about Dr. Fujii finding himself trapped between two beams of his collapsed porch, partly immersed in the rising waters of the Kyo (or Kyobashi) River. He recounted Dr. Sasaki's ordeal at the Red Cross Hospital: the thousands of walking wounded converging upon the building, overwhelming the young surgeon until he became, Hersey wrote, "an automaton, mechanically wiping, daubing, winding, wiping, daubing, winding."

In a third chapter—which Hersey originally titled "With Full Tears in Their Eyes" and then retitled "Details Are Being Investigated"—he recounted the horrific first night that followed the bombing at Asano Park, where several of his protagonists had by then overlapped amidst the aftermath. Here he reported on Reverend Tanimoto's grim ferrying work and the burned skin that slipped off when he tried to lift a blast survivor by the hand into the punt.

When the Americans then bombed Nagasaki, Hersey reported, it would actually be days before Hiroshima's blast survivors knew that another Japanese city had suffered a comparable nuclear attack, for news about the bomb was then still being suppressed by the Japanese government.

THE REAL EFFECTS

In the fourth and final section of his draft, Hersey reported on the extended aftermath of the bomb in Hiroshima. Although his article was primarily built from these six testimonies, Hersey also had to provide information about the atomic bomb and its radioactive aftereffects on Hiroshima's environment and the human body. Hersey had obtained a cache of post-bomb Japanese scientific studies to draw from, including a damage report compiled by the city of Hiroshima, a botanical study on the effect of the bomb on Hiroshima's plants and trees, and a clinical investigation of the "Atomic Bomb Disease" created by a medical clinic at Kyushu Imperial University, which detailed the symptoms and causes of the radiation sickness ravaging survivors. He even had copies of Father Kleinsorge's blood count records. Hersey intended to make it clear that Japanese investigations and findings about the atomic bomb and its aftermath had been concealed by SCAP and officials in Washington, D.C.

"General MacArthur's headquarters systematically censored all mention of the bomb from Japanese scientific publications," he wrote, "but it could not censor the minds of men."

Even though the results of the various investigations had been suppressed, they had since become well-known among Japanese scientists, doctors, and officials, he reported; these communities had known far more than the American population all along, which was intentionally being kept in the dark. He recognized that the suppression of facts about the bomb likely had to do, at least in part, with retaining its nuclear monopoly for as long as possible.

"[But] trying to keep security on atomic fission is as fruitless as trying to keep a blanket of secrecy on the law of gravity," he wrote angrily. "And all the senators and all the generals in America cannot suppress what happened at Hiroshima and Nagasaki any more than they can hide many other things with presumptive military significance, such as the Mississippi River and the Rocky Mountains."

Provisionally titled "The Real Effects," Hersey's fourth and final section depicted the fates of Father Kleinsorge and the other Catholic priests, Reverend Tanimoto, Drs. Fujii and Sasaki, Mrs. Nakamura and her children, and Miss Sasaki in the days, weeks, and months following August 6, 1945. Hersey unsparingly portrayed their unhealing injuries, their poverty, and their struggles with incapacitating radiation sickness. Dr. Fujii, Hersey wrote, seemed to have been spared the worst of the latter affliction; the doctor himself conjectured that his hospital—which had fallen on top of him—may have shielded him from radiation emitted by the bomb. But Father Kleinsorge, Reverend Tanimoto, and Mrs. Nakamura had suffered horribly from various radiation-related symptoms, including fevers, vomiting, and malaise. Within a month of the bomb, Hersey wrote, Mrs. Nakamura was wracked with illness. Swaths of her hair had begun falling out just a couple of weeks after the bombing. Around that time, Father Kleinsorge had become too exhausted to move; his cuts had refused to heal, prompting the rector of his novitiate to accuse him of tampering with the wounds. Why else would they have worsened so?

Even the most aggressive efforts by Hersey's reporter forerunners, like Wilfred Burchett and George Weller, to describe the menace of "Atomic Plague" and "Disease X" had been rendered only in broad strokes. Hersey now intended to reveal in detail the stages of the "atom bomb disease" plaguing blast survivors and show how the weapon continued to kill indefinitely after detonation. He carefully pulled and paraphrased information from the Japanese studies and interviews he had done on the ground. Japanese doctors had quickly concluded, he wrote, that this was a new, man-made illness generated by an assault on the body of neutrons, beta particles, and gamma rays, which decimated and dissolved the body's cells.

The symptoms—as experienced by several of Hersey's protagonists—included nausea, headache, diarrhea, malaise, and fevers that could reach 106 degrees Fahrenheit, followed by sudden hair loss and blood disorders: bleeding gums, plummeting white blood cell counts, infection-prone wounds that wouldn't heal, and purplish-red spots on the skin. The severity of the symptoms appeared directly related to the amount of radiation each victim had received when the bomb detonated. For anyone reading Hersey's story, there would no longer be any question that the atomic bomb was not a conventional weapon, and that radiation sickness was not a "very pleasant way to die," as General Groves had put it.

Also drawing on the Japanese investigations, Hersey reported that blast survivors' reproductive processes had been affected by the radiation: there were reported incidences of bomb-related sterility, miscarriages, halted menstrual cycles—"as if nature were protecting man against his own ingenuity," he concluded.

There was also the question of casualties. Nearly a year after the bombing, still no one knew how many people had died in the Hiroshima blast and its aftermath. Hersey had different estimates to evaluate. One of his sources—a study conducted by the city of Hiroshima—stated that more than 78,000 civilians had died by November 30, 1945, with nearly 14,000 still missing; these statistics did not include members of the Japanese military. However, no one in Hiroshima's government knew for sure how remotely accurate these numbers were, Hersey wrote. Another Japanese report in Hersey's possession cited a figure of 270,000 dead and wounded. (The U.S. government, by that time, had estimated Hiroshima's death toll to be between 70,000 and 80,000 but acknowledged that "the exact number of dead and injured will never be known.") As more bodies continued to be uncovered over the months, Hiroshima officials told Hersey that they were estimating that around 100,000 people had died in the bombing. This is the figure Hersey chose to cite in his story.

Ironically, less than two weeks after Hersey arrived back in America, the U.S. government released a U.S. Strategic Bombing Survey report detailing the United States' own findings on the damage to Japan during its entire wartime bombing campaign. President Harry Truman had requested the study months earlier, on the day of the Japanese surrender. Hersey got a copy of it. The survey's declared purpose: to create a "fairly full account of just what the atomic bombs did at Hiroshima and Nagasaki" and set the record straight about what the survey's authors cited as other, distorted accounts.

Another purpose of the survey, the report frankly admitted, was to study the effects of the atomic bombs on people and urban areas and apply lessons learned to "problems of defense." The U.S. government was already looking ahead to a time when it no longer enjoyed nuclear hegemony and was finding utility in studying Hiroshima's human guinea pigs. "What if the target for the bomb had been an American city?" the survey authors asked. "The danger is real." Luckily, in studying Hiroshima and Nagasaki, the surveyors had learned some lessons that the United States could heed to cut down potential loss of life and property in the event of nuclear attack. For example, too few Hiroshima and Nagasaki residents had had access to underground shelters and been duly incinerated; therefore, the takeaway here was that U.S. cities would need a system of carefully built fallout shelters.

Furthermore, the U.S. Strategic Bombing Survey report went on, the fate of the atomic cities had demonstrated the "value of decentralization" in the atomic age. Because so many of Hiroshima's medical facilities had been based in the city center, they had been "crippled or wiped out by the explosion." American urban planners might now want to consider "wise zoning" and a "reshaping and partial dispersal of the national centers of activity" lest the same fate befall U.S. cities and citizens. The experience of the people of Hiroshima and Nagasaki had been most helpful in terms of making these new needs apparent. (The report's authors also admitted that "our understanding of radiation casualties is not complete"; further studies of the blast survivors would be needed—and were indeed forthcoming.)

Upon reading the U.S. Strategic Bombing Survey report, Hersey grew even more incensed and poured his frustration into his article. He reported that the government was still withholding information, such as height of the bomb burst and the amount of uranium used; he also added that additional, secret parts of the report had not been released to the public.

The closing part of Hersey's story, however, belonged to his protagonists and the challenge of survival and starting anew in their demolished world. Nearly a year later, the six had varying attitudes toward the bomb and the Americans. Mrs. Nakamura and Dr. Fujii had become resigned. Father Kleinsorge and the priests were still debating the ethics of the bomb's use. Dr. Sasaki was far less philosophical about it: he told Hersey that the Americans who had dropped the bomb should, like the Japanese high command currently on trial in Tokyo, also be tried for war crimes and hanged.

Initially, Reverend Tanimoto was given the last word in the article as he mulled over other, possibly positive uses for atomic energy. This did not ultimately strike the darkly poignant note Hersey desired, for he would eventually rewrite the story's last sentences, giving the final word to Toshio Nakamura, Mrs. Nakamura's young son, who was ten years old at the time of the bombing. Hersey recounted Toshio's childlike recollections of the fateful event: The day before, he'd been eating peanuts. After the strange light had flashed, his neighbors were bloody. He and his family escaped to the park, where they saw the whirlwind. He came across two of his young friends, who were looking for their mothers.

"But Kikuki's mother was wounded," Hersey wrote, "and Muraka-mi's mother, alas, was dead."

IT'S GOT TO RUN ALL AT ONCE

Hersey's article sprawled to around 30,000 words. The *New Yorker*'s editors had originally planned to run the story in several installments as a "A Reporter at Large" serial. But upon reviewing it all, Shawn realized quickly that the serialization plan was a no-go.

"Look, we just can't," he told Hersey.

Keeping track of the ensemble of protagonists from installment to installment was too complicated; the story would lose its pace and impact. Squandering this important reporting in that way would be unacceptable, Shawn felt. The editor was clearly as ambitious for the story as Hersey was—if not more so. Rather than having Hersey cut the article down into a shorter, less monumental report, Shawn saw an opportunity to do something that would actually make the article even more provocative.

"This can't be serialized," he told Harold Ross. "It's got to run all at once."

It would easily be the longest single-issue article that had ever run in the *New Yorker*. Furthermore, Shawn told Ross, he should consider excluding all of the other material in the issue featuring Hersey's story: the "Talk of the Town" pieces, the fiction, the other articles and profiles—and, of course, the urbane cartoons, which would seem the height of crassness when juxtaposed with details about Hiroshima's charred corpses.

Ross was taken aback by the suggestion, which would amount to "an unprecedented editorial splurge in his or any other magazine," as one longtime *New Yorker* contributor later put it. At first Ross protested that Hersey's story, if run all at once, might prove too jarring. Yes, they wanted to shake things up with this story, but Ross worried that their peacetime readership might not be prepared for such an undiluted dose of atrocity reporting. Plus, he added, they might feel "cheated" out of their favorite regular features. Even during the darkest years of the war, after all, the *New Yorker* had run its cartoons, its "Goings on About Town" section, and its "Talk of the Town" items.

The proposal prompted an editorial existential crisis for Ross akin to the one he had experienced over the magazine's identity and direction in the lead-up to the war. But the attack on Pearl Harbor had resolved any qualms Ross had had about turning the *New Yorker* from a humor maga- zine into a platform for serious conflict journalism.

Since the Japanese surrender, Ross had been struggling with how— and whether—to resume the magazine's amusing pre-war tenor after "having gone heavyweight to a considerable extent during the war," as he put it. He thought the magazine's war reporting had been essential and stupendous. Now he had the opportunity to run one of the biggest and most essential stories of the war, and roll it out spectacularly. But the Hiroshima story was already going to be extremely controversial even if they ran it over several issues; running it in a single issue would be akin to putting movie premiere searchlights on it. The one-issue question was testing the editors' resolve in advance. Ross fretted about the situation to longtime *New Yorker* writer and editor E. B. White.

"Hersey has written 30,000 words on the bombing of Hiroshima (which I can now pronounce in a new and fancy way)," he told White. "[It's] one hell of a story, and we are wondering what to do about it." Shawn was pushing hard to print the story and nothing else in one issue, Ross said: "He wants to wake people up and says we are the people with a chance to do it, and probably the only people that will do it."

After torturing himself about the decision for the better part of a week—during which time Shawn continued to lobby for the cause in his gently relentless way—Ross decided to seek guidance from his maga- zine's own DNA. He reached for the very first issue of the *New Yorker*, published on February 21, 1925, and read on pages one and two, the magazine's original statement of purpose—the one in which Ross had famously impugned the "old lady in Dubuque."

The *New Yorker*, he had written twenty-one years earlier, was going to be "gay, humorous, [and] satirical." But he also announced that the magazine was starting off "with a declaration of serious purpose" and that it would "publish facts that it will have to go behind the scenes to get." It also pledged to "try conscientiously to keep its readers informed." The 1924 *New Yorker* prospectus that Ross had written to announce the imminent publication of the magazine, and from which he'd drawn the language of the first issue's statement of purpose, further added that the *New Yorker* would "present the truth and the whole truth without fear." This was exactly what Ross needed to convince him that Shawn was right. He called Shawn and Hersey at their homes and informed them that they would indeed be devoting an entire issue to Hersey's full account.

A few weeks later, in confidence to *New Yorker* writer Rebecca West, Ross made the agonizing decision-making process sound more cavalier. "After a couple of days more of reflection," he said, "we got into an evangelical mood and decided to throw out all the other text in the issue, and make a gesture that might impress people." The Hiroshima issue, he added, was certainly going to be distinctive.

"I don't know what people will think," he told West, "but a lot of readers are going to be startled."

THE GREATEST STRAIN EVER IMPOSED ON A MAGAZINE STORY

Over the next ten days Hersey and the two editors sequestered themselves in Ross's corner office to edit the story. It was an austere space, adorned almost solely by dictionaries, a bare, smoldering radiator, and a tilting hat rack. Other flourishes usually included Ross's sloppy briefcase, a bottle of his ulcer medicine, and a copy of *Who's Who*. Now the desk and table were covered with pages of rewrites of Hersey's magnum opus. Starting each day at 10:00 a.m., the team edited and rewrote until 2:00 a.m. under the gaze of unpublished James Thurber drawings staring down from Ross's walls.

Ross and Shawn had decided that the Hersey project must be kept top secret—even from the in-house *New Yorker* team. With the exception of Ross's secretary and perhaps another secretarial assistant, and the magazine's senior layout man and production manager, no one at the *New Yorker* was told about the explosive single-story issue the Hersey team was preparing. The project was their own journalistic version of the Manhattan Project. The staff and contributors knew that some major, clandestine editorial operation was going down behind Ross's locked office door—the *New Yorker* headquarters was tiny, after all, and secrets could be kept only so secret—but no one knew exactly what. (They would all be "gobsmacked" when they found out weeks later, said one Ross biographer.)

As Hersey, Shawn, and Ross worked furiously on the story, much of the rest of the *New Yorker*'s team was reportedly kept busy working on a "dummy" issue that they *thought* would be going to press. The hoodwinking was necessary, the editors felt. "There would have been no way to keep it quiet otherwise," recalled Ross biographer Thomas Kunkel. "At the *New Yorker*, they were always working on several issues at one time: the A issue, the B issue, the C issue, and so on; if one of them had just evaporated, everyone would have known." Stories and copy were being approved but secretly stashed aside for later; contributors were baffled and eventually disgruntled that no layout proofs containing their stories and artwork materialized.

Nor was the *New Yorker*'s business office let in on the secret. The staff there just assumed that the weekly advertisements would run alongside the usual cartoons, short pieces, and features. Advertisers were, by extension, also kept in the dark: the makers of Chesterfield cigarettes,

Perma-Lift brassieres, Lux toilet soap, and Old Overholt rye whiskey would just have to find out along with everyone else in the world that their ads would be run alongside Hersey's grisly story of nuclear apocalypse. Maintaining the big subterfuge may have been difficult for Ross but easier for Shawn. "Ross could never keep a secret for long," recalled longtime *New Yorker* writer Brendan Gill, "but Shawn relishes keeping them and will go to his grave in happy possession of many thousands of confidences."

Ross had liked Hersey's first draft. It was, in his estimation, "a very fine piece beyond any question" and had "practically everything." He felt it was poised to become the definitive piece on the dropping of the bomb. That said, he prepared an onslaught of queries and edits. There were scores of Rossian notes in the margins of the proofs. Even Ross admitted that he may have "read it over-zealously."

The article's title—at that stage it was still being called "Some Events at Hiroshima"—was dissatisfying to the founding editor and needed to be changed. Ross also thought Hersey needed to add in some sort of breakdown of how the 100,000 people had died: "How many were killed by being hit by hard objects, how many by burns, how many by concussion, or shock, or whatever it was." (Shawn may have struggled with the story's more ghoulish medical passages: he apparently had "a strong aversion to clinical details that concern the human body, especially when they concern blood," according to Brendan Gill, who added that Ross, on the other hand, as "a measure of his radical unlikeness to [Shawn,] enjoyed reading and talking about diseases.")

Ross also pointed out that the reader lost all sense of time while reading the article: Hersey would need to plug in the hour or minute from time to time so the reader could orient himself. There was also "trouble with these Jap quotes." Hersey needed to follow each Japanese word he cited with its English translation. In addition, Ross demanded an even greater level of detail from Hersey.

These were just some of the overarching edits. The specifics got very specific. In part one, Hersey had written that one of the German priests had taken refuge during the blast in a strong doorway. Ross's response: "I don't see how a doorway can be said to have strength; a doorway is a hole, a space." Hersey clearly meant a door*frame*, Ross went on—and he ought to know: after all, "as an old air raid and San Francisco earthquake veteran, I know that the idea is that you stand in a doorway and that nothing can fall straight down on you." He objected to Hersey referring to the pole with which Reverend Tanimoto had steered his boat as "slender." First of all, "could one row a boat with a bamboo pole[?]" Ross asked Hersey. He believed Hersey when "he said this Jap did it, but I wouldn't trifle with my luck by making it a *slender* pole." Ross also couldn't resist some commentary on Hersey's story and characters. He was incredulous upon learning of the Jesuit priests' leisurely morning activities just before the bomb went off.

"I'll be damned if it isn't amazing that . . . these Society of Jesus gents . . . all went back to bed after breakfast," he wrote. "I've been suspicious of these religious men all along, and envious. Ah to be a member of the Society of Jesus."

The editing even kept Ross awake at night. He was tormented in the early hours by a sentence about a pile of tangled bicycles, which Hersey had described as having been made "lopsided" in the bombing. He subsequently regaled Hersey and Shawn on the subject: How could something that was two-dimensional be lopsided? Ross demanded to know. Hersey and Shawn both went to their homes that night and tried to think of an alternative word. Hersey settled on "crumpled." When he arrived at the *New Yorker* offices the next morning, Shawn had already written the same word down on the proof. For Hersey, this was further evidence of Shawn's editing ESP. He was, Hersey believed, "a kind of editorial Zelig" who had the ability "to think in the vocabulary of the writer he's editing."

Even if Shawn was the resident editorial diplomat, he was equally unremitting in the quest for accuracy. Future generations of the magazine's writers would see in Hersey's story all of the hallmarks of Shawn's editing. "If you're from the *New Yorker*, you can tell where Shawn had gone over it heavily," says longtime *New Yorker* staff writer Adam Gopnik. "He had an understated, empirical, punctiliousness combined with a certain kind of sober moral indignation."

The team missed the August 6 anniversary. The article was now scheduled to take up the entire August 31 issue. They needn't have worried about competitive anniversary reporting. On August 7 the *New York Times* ran, on its thirteenth page, a short item about Hiroshima titled, "Japan Notes Atom Anniversary; Hiroshima Holds Civic Festival." (In the article, *Times* reporter Lindesay Parrott informed readers that "few signs have as yet been found of permanent injury caused by rays to those who survived," citing a reportedly just-released and unnamed Japanese survey.) The editors at *Time* magazine waited nearly two weeks before running a minuscule anniversary item headlined, "Japan: A Time to Dance." The unsigned story noted that "thousands of [Hiroshima] citizens carried on as if the first anniversary of the atomic bomb were Rodeo Day in the Texas Panhandle." They had, *Time* reported, flocked to movie houses, held a "bangup ritual lantern dance" at a Shinto shrine, and "stampeded in the city's makeshift department stores to take advantage of bargain sales in Hiroshima-made products."

Gradually, painstakingly, Ross, Shawn, and Hersey tightened the story. Hersey's draft copy was tweaked into final form, including an updated version of the work's soon-to-be most famous lines: of Miss Sasaki's ordeal, he rewrote, "There, in the tin factory, in the first moment of the atomic age, a human being was crushed by books." Mrs. Nakamura's son was given the final say in the article instead of Reverend Tanimoto's subdued musings about the peaceable applications of atomic energy. The

100,000 death toll figure was added to the first section, to remind readers immediately of the extent of the wrath inflicted by a single, primitive nuclear bomb.

Despite the glib tenor of some of Ross's correspondence during the edits, the team had given the story an intense going-over, even by the strict standards of the magazine. They needed to tear apart the article, to analyze it from every angle, and scrutinize every word because, as Ross told Hersey, it was poised to become "the most sensational one of the generation." The *New Yorker* had as much at stake in the story's publication as Hersey. Factual or editorial missteps would be disastrous for both parties.

"Hiroshima," as the team had retitled the story, was about to become a landmark in the magazine's history—and could become a turning point in the nation's history as well. Most Americans still wholeheartedly approved of the use of the atomic bombs on Japan; they continued to luxuriate in what they saw as America's morally unqualified victory and still held sacred the idea that the Japanese had had it coming. They had little or no idea what it had been like to suffer the nuclear attacks, nor did they understand the longstanding effects of these still-experimental weapons.

In Hersey's "Hiroshima," Americans were about to be confronted with the realities of the wrathful, Godlike military actions conducted covertly in their name, and what the future of warfare might look like. Although the article never directly questioned the use of the atomic bombs, it would inevitably swerve a glaring spotlight back onto those who had created and used them—from President Truman to Oppenheimer to General Groves—and it would certainly expose the extent to which those principals had covered up the less savory aspects of their handiwork.

Therefore, "Hiroshima" had to be flawless. After all, as Ross told Hersey, it was "going to bear the greatest strain ever imposed on a magazine story."

RESTRICTED DATA

During the war, the editors at the *New Yorker* had submitted their war stories for clearance to the War Department along with everyone else. They rarely if ever got any significant blowback or changes from the War Department's public relations team; correspondences among the magazine's editors and censors were cordial. Submitted articles were usually returned by the War Department's press officers to the editors promptly with desired changes noted, or else approved with a "no objection to publication" designation.

An executive order officially terminating the wartime Office of Censorship had been signed on September 28, 1945. Yet, in the months that followed, the *New Yorker*'s editors had continued to submit for clearance story drafts that touched on nuclear matters to comply with the government's confidential order issued that previous fall, requiring that such material be submitted to the War Department review in the interest of national security. Just a few months earlier, in May 1946—while Hersey was still waiting in China for SCAP clearance to travel to Japan—the *New Yorker* had submitted to the War Department a proposed story by correspondent Daniel Lang, in which he had interviewed Dr. Philip Morrison, a Manhattan Project physicist who had accompanied General Thomas Farrell on his "spot check" of Hiroshima the previous summer. The *New Yorker* editor who submitted the story thanked the public relations officer for his expedition "in the matter of censoring" and for "this and past favors you have done us." The story was quickly cleared and ran when Hersey was still on the ground in Hiroshima.

But then—as Ross, Shawn, and Hersey were hunkered down, editing "Hiroshima" in the *New Yorker* offices—something happened that created for the team a significantly more perilous legal landscape: on August 1, President Truman signed into law the Atomic Energy Act. Among its other provisions, the act established a "restricted data" standard that included "all data concerning the manufacture or utilization of atomic weapons, the production of fissionable material, or the use of fissionable material in the production of power." Anyone who had access to what was deemed restricted data—whether he had gotten that information lawfully or not—and communicated, transmitted, or disseminated that data "with any reason to believe that such data" could be used to harm the United States could face imprisonment and substantial fines. If it could be proved that that the individual was actively attempting or conspiring "to injure the United States" or to "secure an advantage to any foreign nation," he or she could even "be punished by death or imprisonment for life."

In "Hiroshima," Hersey had steered clear of material that had previously been cited as classified by the U.S. government: the height of the bomb's detonation, the size of the resulting fireball, and so on. (He did, of course, point out in his draft that such information was being withheld from the American public.) That this information was restricted was now well-known throughout the industry; the specifics had been detailed to editors in the past year and had also been included in list form in the guidelines distributed to the journalists witnessing the Bikini tests.

However, the Atomic Energy Act did not provide a specific list of what information might now be considered restricted. Ross contacted their lawyer, Milton Greenstein, the day that President Truman signed the act into law.

"Should we submit ["Hiroshima"] to censorship?" he asked. "Mr. Shawn and I don't want to, but we don't know whether the law is that we shall do so." He added that all of Hersey's information "has come from Japanese sources"—which was not strictly accurate—and that "the Army supplied none of it." Could Greenstein "find out where we stand, please?"

The lawyer evaluated the new law and Hersey's manuscript.

"'Data' is not defined," he confirmed to the editors, "but I believe that as used in the act it refers to scientific and technical matters." He did not believe that anything in "Hiroshima" could be considered restricted, and, "of course, we are not publishing anything in the article 'with intent to injure the United States.'" That said, Greenstein added that "there may be a few observations reported by Hersey which *might* be called scientific." If they were in doubt about whether some information could fall into the restricted category, he continued, they "probably ought not to 'disseminate' it."

The *New Yorker* team now found itself in a hell of a quandary. They could defang the story or kill it entirely, or they could run it as it was and risk severe legal penalties. If "Hiroshima" indeed contained information that the U.S. government might now be able to classify as restricted, prosecutors would still have to prove that Hersey and the *New Yorker* intended to harm the United States or aid its enemies—a difficult case to make. Yet, with this new act, the government would be in a stronger position to claim that the *New Yorker* had knowingly put the country at risk by revealing dangerous information, resulting in a public relations disaster for the magazine at the very least. Even if the team couldn't be successfully prosecuted for any alleged transgressions, there could be a backlash against and boycott of the magazine and its advertisers. The consequences could "even be enough to close the publication," as one censorship historian put it.

How the *New Yorker* team ultimately arrived at its decision about submitting the story for government vetting is unclear, but at some point in that first week of August, Ross and Shawn made what must have been a painful choice. They submitted "Hiroshima" for review to the War Department—and not just to any public relations officer there, but to General Leslie Groves himself.

The editors did not disclose to General Groves their plan to showcase "Hiroshima" so dramatically in a single issue. The submission appears to have been casually presented, as if it were just another in a long line of war-related *New Yorker* stories. Shawn simply categorized it as a "four-part article on the bombing of Hiroshima" to the general.

The team waited for his reply.

CHANGING THE ARTICLE A LITTLE

At first glance, the submission of the "Hiroshima" manuscript to the War Department for clearance may seem akin to sending Hersey's magnum opus straight to the guillotine. Yet, if Ross and Shawn decided that government clearance of the story had become a necessary evil, General Groves might have actually been an unlikely loophole, and the submission a calculated gamble on the editors' part.

Even though General Groves had tried in the early months to cover up the full effects of the bomb—especially radiation sickness—a year later new considerations and circumstances had apparently altered the general's attitude toward public perception of the bombs. While at first he had tried to depict the bomb as relatively humane, his lack of repentance about its severity had become a matter of public record. "As I look at the pictures of our men coming home from Japanese prisons and hear second-hand accounts and first-hand accounts of the experiences of the men who made the march from Bataan, I am not particularly worried about how hard this weapon hit the Japanese," he stated.

The *New Yorker* editors might have reasoned that even if General Groves ended up excising some of the draft's technical information about the bombing in the professed interest of maintaining national security, perhaps the rest of the story—the experiences of Hersey's six protagonists—had a shot at remaining intact, given the general's indifference to the suffering of his former adversaries. From a perverse point of view, the eyewitness accounts within "Hiroshima" could even be seen as an advertisement for the effectiveness of the weapon whose creation General

Groves had spearheaded—and he had become increasingly concerned with receiving credit for his role in creating the war-winning weapon.

The *New Yorker* editors and Hersey also knew from the U.S. Strategic Bombing Survey report that the U.S. government was now starting to see utility in studying Hiroshima's bombing victims, as their experience during the bombing and in the aftermath was starting to prove instructive in preparing the U.S. military, the government, and medical researchers for a possible nuclear attack against America someday. General Groves felt that the United States needed to build out its nuclear arsenal. The United States could only hold on to its atomic monopoly for so long; even though General Groves believed that the Soviets—now America's great Cold War adversary—might still be five to twenty years away from getting their own atomic bomb, he was already seeing the need to seek public support for creating mightier versions and greater stockpiles of nuclear weapons.

"If there was only some way to make America sense now its true peril some 15 or 20 years hence in a world of unrestricted atomic bombs," he had written in a memo earlier that year. Atomic weapons were now a permanent part of the world, and, he concluded, "we must have the best, the biggest and the most."

As one of the general's emerging goals was to generate public support for the United States maintaining its nuclear superiority, perhaps Ross and Shawn reasoned that General Groves might see a story like "Hiroshima" as an asset to help him make his case. If readers were able to imagine their hometowns of Houston or Akron or New York in Hiroshima's stead—as Hersey, Shawn, and Ross had hoped—they might indeed cry out for a ban on nuclear weapons and revile the figures who'd conjured them into existence in the first place. Or, instead, they might suddenly start to see an urgent need for the United States to maintain its nuclear supremacy and call for the country to add rapidly to its nuclear arsenal, as General Groves hoped. (And until the Soviets did get the bomb themselves, "Hiroshima" might also serve to remind them of the terrible disadvantage at which they currently found themselves. In this respect, "Hiroshima" might certainly serve as a handy bit of PR for the United States.)

At 3:20 p.m. on August 7, the general called Shawn at the *New Yorker*'s offices. He was greenlighting the story, he told the editor. However, the general continued, he wanted to discuss "changing the article a little"—just a couple of alterations in different places—adding that his edits "would not hurt the article." Could he send one of his public relations officers to the *New Yorker* to discuss the changes in person?

Shawn agreed. Arrangements were made for one of General Groves's public relations officers to come to the *New Yorker* the next morning. The details of the subsequent meeting are unknown. Neither the *New Yorker*'s surviving records, nor Hersey's records, nor General Groves's records indicate exactly which information the general wanted cut or changed. Dated, incremental drafts of Hersey's story are missing from the *New Yorker*'s archive and his own papers at Yale, and no documents indicating the desired changes appear to exist among the general's records. Yet certain contentious parts of the first draft of "Hiroshima," which has survived, did not make it into the published version.

In the final, post–War Department meeting version of the article, gone was Hersey's line about Americans being willfully kept in the dark about the exact height of the bomb's detonation and the weight of uranium used; gone was his indignant line that "trying to keep security on atomic fission is as fruitless as trying to keep a blanket of secrecy on the law of gravity." (Probably predictably, also cut was Hersey's original line that "all the senators and all the generals in America cannot suppress what happened at Hiroshima and Nagasaki.") No longer did the article criticize the U.S. Strategic Bombing Survey report; nor did it state any longer that classified, unpublished parts of that report existed. However, some new lines had made their way in, including the fact that a bomb "ten times as powerful—or twenty—could be developed."

Some other surprising elements stayed in, such as Hersey's line about General MacArthur systematically suppressing mention of the bomb in Japanese publications. (Perhaps this wasn't so surprising after all: there was little love lost between General Groves and General MacArthur.) Hersey's depictions of the radiation illness plaguing several of his protagonists remained intact in the published version as well, but this was because the article implied that the patients had incurred their radiation exposure at the moment of the blast, instead of via residual radiation— still an extremely touchy subject for General Groves and the U.S. government, which had been at pains to deny the existence of lingering radioactivity in the atomic cities. (Plus, the public U.S. Strategic Bombing Survey report had publicly acknowledged that radiation emitted from the bomb had also killed some blast victims, perhaps paving the way for lenience when it came to Hersey's account.)

If General Groves had been bothered by Hersey's disturbing descriptions of the Japanese soldiers with their liquefied eyeballs streaming down their faces, he didn't ask for that passage—or similarly ghoulish ones—to be stricken. After all, he—like President Truman—believed that the Japanese had simply been repaid in kind.

Shawn submitted another version of "Hiroshima" to General Groves "to look over" on August 15 and made an entreaty for final input the following day. Apparently he got it—along with the general's blessing—for the August 31, 1946, issue went into final production. Luckily for him and the *New Yorker* team, General Groves appeared to have overlooked Hersey's most unsettling revelations: the fact that the United States had unleashed destruction and suffering upon a largely civilian population on a scale unprecedented in human history and then tried to cover up the human cost of its new weapon. Whatever concessions the *New Yorker*'s editorial team had had to make, for them, "Hiroshima" had survived the review process to become a document of conscience and an urgent warning about the future of civilization in the atomic age.

PRIDE—AND HOPE

"Hiroshima" had survived its near-death experience largely unscathed. The essence of what Hersey and Shawn had originally wanted to achieve—a subversive, harrowing account of the bomb seen through its victims' eyes—had been preserved. Now that the *New Yorker* editors had the almost final version in hand, they prepared to present it to the world.

The magazine's covers were usually selected and assigned months in advance. The one slated to adorn the August 31 issue was an illustration by artist Charles E. Martin, who had, during the war, worked for the Office of War Information, creating leaflets to be dropped behind enemy lines. For this August cover, however, Martin had created a carefree scene in an unnamed park in which sunbathers reclined next to a lake, smiling, eyes closed; revelers played golf, croquet, and tennis, and rode horses and bicycles; and one gentlemen fished contentedly while smoking a pipe. Here was America at leisure again.

The editors chose to keep Martin's summery picture on the "Hiroshima" issue. After reading the story inside, this bucolic scene would likely take on unnerving connotations for readers: perhaps this was a sleepwalking, apathetic America that had indeed "escape[d] into easy comforts," as Albert Einstein had put it just months earlier, while ignoring the perils of the atomic age. Or it could have been seen as echoing the obliviousness of Hiroshima's residents as they went about their daily routines on the morning of August 6, 1945, just before 8:15 a.m. (Some of the story's readers might have found the cover scene especially sinister after realizing that a comparable public park in Hersey's story—Asano Park—had ended up serving as an evacuation site for blast survivors, and within twenty-four hours was littered with scalded corpses.)

The editors did, however, grow concerned about the lack of advance warning to readers about the graphic contents of the magazine. "My God, how would a guy feel, buying the magazine intending to sit in a barber's chair and read it!" someone brought up. (At that time, *New Yorker* covers did not have text describing the issues' contents; nor did the magazine have a table of contents inside.) Some sort of visual alert would have to be quickly conjured up. Ross was able to order white paper bands that would encircle the 40,000 New York newsstand copies with a warning that the issue contained a disturbing story.

Other than the cover illustration, the white paper magazine belt, and a small, hand-drawn, map-like illustration of Hiroshima depicting its fan-like shape intercut by rivers, "Hiroshima" would be adorned by no visuals. (The team had attempted an illustration of a mushroom cloud to accompany the story, but it was deemed a distracting failure.) Previously released Hiroshima photographs had failed to convey the horror there. The imagery would come instead from Hersey's words. Photographs of the six survivors—taken by Hersey—would be made available to other publications via Acme Newspictures, Inc., a photo agency. Hersey had agreed that the *New Yorker* could allow other publications to reprint "Hiroshima," but only if those publications agreed to run the story in its entirety. Furthermore, he made it clear to the team that he did not want to benefit financially from any income resulting from syndication.

"Like other Americans, I felt some guilt about the bomb and about profiteering on it, and I decided to give away the income from the first re-prints," he later said.

It was decided that he would give reprint proceeds to the American Red Cross. In a conference call, the magazine's treasurer, R. Hawley Truax—now in on the "Hiroshima" project—suggested that the team advertise this donation. Hersey, Ross, and Shawn agreed.

As the project was still largely under wraps within the *New Yorker* office, Shawn and Ross personally ferried the article proofs by train to their press in Connecticut.

"I feel as weak as if my aunt had run away with the ice man," Ross said. "And of course as proud."

Shawn, as usual, was more reverential. Upon delivering to Hersey a final mock-up of the issue, he included a note.

"Dear John," he wrote. "A rough copy of the issue, which I send to you with gratitude and unlimited admiration—and with hope."

Chapter Six

Detonation

THIS IS THE STORY OF WHAT
WOULD HAPPEN TO YOU

On the morning of Thursday, August 29, 1946, tens of thousands of copies of the *New Yorker* arrived at newsstands, landed on welcome mats, and nestled in mailboxes across the country. Readers would have the long Labor Day weekend to read and react to "Hiroshima."

The morning of the issue's release, Lillian Ross, then a new reporter at the *New Yorker*, was summoned to William Shawn's small spare office. His desk was, she recalled later, a wooden table topped with tidily stacked long galley proofs and a cup filled with sharpened black pencils. Looking tense, the editor asked Ross to go to Grand Central Terminal and see if people were lining up at the newsstands to buy the white-banded issue.

"I hurried over there," Ross recalled later, "and found no line, no crowd. I returned to Bill and hesitantly gave him my report."

Shawn's disappointment was palpable. "I thought that the entire town would be in an uproar," he lamented. "I thought they would be paying attention."

His dismay would not last long. As the day progressed, the reaction turned from tepid to "explosive," as Hersey later stated. Readers opening the issue and beholding the article were greeted by a short editor's note in boldface on the first page of "Hiroshima":

> *TO OUR READERS:* The New Yorker *this week devotes its entire editorial space to an article on the almost complete obliteration of a city by one atomic bomb, and what happened to the people of that city. It does so in the conviction that few of us have yet comprehended the all but incredible destructive power of this weapon, and that everyone might well take time to consider the terrible implications of its use.*
>
> **—THE EDITORS**

Word of mouth about the story was bound to be considerable, but Ross and Shawn were leaving nothing to chance. After weeks of secrecy, it was time for the big reveal. The day before, the editors had dispatched copies of the story to editors at nine major New York newspapers and three international wire services. Along with the article, they had included a letter from the editors stating that the magazine had "broken a precedent of twenty-one years standing" in devoting the issue solely to one story, adding that Hersey's article was "a terrifically important one."

In case any of the recipients was inclined to dismiss a story on Hiroshima as old hat, Ross and Shawn highlighted several of Hersey's newsy findings, including the recently reported casualty count of 100,000 dead, the apparent silence of the explosion, and how the Japanese scientists had first figured out what sort of weapon had hit the city. The *New Yorker* editors did not mention, however, that the story would be the first to depict the Japanese victims as ordinary human beings—or human beings at all—a then-revolutionary approach to the subject of the atomic bombings; nor did they call attention to how Hersey had revealed that the true story of the Hiroshima bombing had long been covered up. Their editor colleagues around the world would glean this on their own.

Anxiety levels were high at the *New Yorker* offices in the lead-up to the story's release. Ross later told another editor that the "Hiroshima" team had "gone out confidently," but that he privately worried at the eleventh hour that they may have "gone out on the limb" after all. Hersey left New York City altogether after finishing the proofs, heading to the tiny town of Blowing Rock, North Carolina, at the crest of the Blue Ridge Mountains. Perhaps he anticipated outsize blowback from the article's release; his reasons for departing have not been recorded, although the action was consistent with what became his lifelong refusal to publicize his own works. His absence on the eve of the release of "Hiroshima"—poised to become the most controversial work of whistleblowing journalism in recent memory—was met with confusion by some members of the press. Some were disdainful of the move (Hersey had "ducked out of town," reported *Newsweek*); others were a little more understanding (the response to the article was bound to be so overwhelming that Hersey had had to leave, explained another publication). In any case, Blowing Rock was a somewhat ironic choice of outpost, given Hersey's subject matter: the town was known for its omnipresent winds, which often blew vertically upward and could be strong enough to catapult objects straight up into the sky.

The news organizations that received early copies of "Hiroshima" immediately took the bait; several raced to get out the first story about Hersey's reporting coup. The *New York Herald Tribune* won. "We managed to blaze the Hersey trail first," one proud *Herald Tribune* editor informed Ross. The paper ran three immediate stories on "Hiroshima," starting the morning of the issue's release with a fervent story from *Herald Tribune* columnist Lewis Gannett.

"Hiroshima" was the best reporting to come out of the war, Gannett announced, and it was about to commandeer the national conversation about what had happened in Hiroshima and about nuclear arms. Everyone would be talking about it for a long time, he added—even those who didn't read it. And those who did read it would never forget it.

"You smell the city of the dead," he wrote. "You live not so much in agony as in stunned bewilderment."

The world had indeed been dazed, exhausted, and sickened after experiencing the horrors of the war, the *Tribune*'s editorial board added in a separate editorial, which had made it nearly impossible when the bombings happened to impress upon humanity the true scope and implications of the new weapon. Plus, the editorial continued, the "old paradox which renders man capable to being deeply moved by the sufferings of individuals, but leaves him dulled by suffering in the mass, has hitherto blocked an appreciation of atomic terror." But finally, the editorial continued, Hersey had brought home the truth, making "the tragedy of Hiroshima real as nothing else published . . . has done."

Scores of newspapers and publications from across the country immediately contacted the *New Yorker* with reprint and interview requests. Editors from more than thirty states quickly asked to excerpt "Hiroshima" or run it in its entirety. (Among those publications was an Albuquerque newspaper, for whom the story was close to home: its office stood just over a hundred miles north of the Los Alamos atomic test bomb site.) "We are hearing from all over the world, and Christ knows what," Ross reported to Gannett. His gamble was paying off: he now had definitive reason to believe that "Hiroshima" would become the most widely reprinted work of journalism of his time.

Even publications unable to reproduce the whole 30,000-word story began running front-page banner stories and urgent editorials about its revelations. The coverage quickly grew so intense that it almost seemed as though Hiroshima's bombing had happened the day before, not a year earlier. Editors reminded readers again and again that Hersey's story could easily have unfolded anywhere in the USA, and that the six survivors could just as well be residents of Cleveland or San Francisco.

"It is what might, and probably would, happen to you and millions of other civilians if there were another war," stated an editorial in the *Indianapolis News*, which ran coverage of Hersey's story with a banner headline blaring, "HIROSHIMA—DEATH OF A CITY."

Editors and columnists across the country also now suddenly decried the silence and secrecy that had shrouded the nuclear aftermath in Hiroshima and Nagasaki in the wake of the bombs. "This article is the first attempt to inform the world of men what really happened there," stated another editorial in California's *Monterey Peninsula Herald*. "Hiroshima," it continued, had clearly revealed that there had been elaborate efforts made by government "to conceal from the American people the full story." After the war, the Germans had professed that they hadn't known what happened in their concentration camps, the editorial pointed out; Americans were now in a comparable position and looked like "amoral fools." Hiroshima had not been treated as a crime because it was a victor's handiwork. Americans immediately needed to be told everything that had happened in this tremendous event and could "tolerate no [more] concealment." The country's moral stature was at stake.

HEADIER THAN HELL

At lunchtime on publication day, Ross got a call from a *New York Times* editor who breezily informed him that the *New Yorker* team had done a "swell job" with the Hiroshima story. That same day the *Times* had run a small item about the "Hiroshima" issue that merely informed readers that the *New Yorker* had devoted its entire issue to Hersey's story and noted that the magazine's famous cartoons were conspicuously absent.

Yet if Ross and Shawn worried that the *New York Times* would handle "Hiroshima" and its significance glibly, the next day the *Times* editors published an astonishing, solemn editorial about the article. Hersey's story appeared to have deeply jolted editors there.

"Every American who has permitted himself to make jokes about atom bombs, or who has come to regard them as just one sensational phenomenon that can now be accepted as part of civilization, like the airplane and the gasoline engine . . . ought to read Mr. Hersey," the editorial read. The *Times* editorial board then took issue with the decision to drop the bombs in the first place. "The disasters at Hiroshima and Nagasaki were our handiwork," the *Times* editorial stated. "They were defended then, and are defended now, by the argument that they saved more lives than they took—more lives of Japanese as well as more lives of Americans. The argument may be sound or it may be unsound. One may think it sound when he recalls Tarawa, Iwo Jima, or Okinawa. One may think it unsound when he reads Mr. Hersey."

Now that Hersey had revealed the catastrophic reality of atomic bombs, the *Times* asked, could Americans ever drop another? If individual readers were still on the fence about this, the editorial continued, they should read Hersey's article, which was not just about the death and destruction of people and cities, but of the human conscience itself.

"History is history," the *Times* concluded. "It cannot be undone. [But] the future is still ours to help make."

This particular editorial was the stuff of nightmares for the U.S. government and military, a high-level public rebuke to their monthslong campaign to characterize the Japan bombs as humane and life-saving. Even though Hersey had never directly questioned the arguments behind the use of the bombs, the *Times* editorial showed that "Hiroshima" had made the first true chink in the government's contention that the bombs had been necessary. The *New York Times* had, up to that point, served as a largely reliable wartime ally for the government, especially given the amount of space the newspaper had devoted to "Atomic Bill" Laurence's War Department–sanctioned reporting on the bomb over the past year.

That the *Times* was essentially treating "Hiroshima" as a bolt from the blue was equally astonishing, given that Atomic Bill had been the sole media witness to the bombing of Nagasaki and the Manhattan Project's in-house historian, and that another *Times* correspondent—"Non-Atomic" Bill Lawrence—had been among the first Western reporters on the ground in *both* atomic cities. Furthermore, the *Times* had maintained a Tokyo bureau since the occupation had begun a year earlier.

It was becoming clear that Hersey, Ross, and Shawn had truly pulled off the scoop-in-plain-sight. The "flock of other journalists" had squandered bountiful opportunities when it came to reporting on Hiroshima. Now their own shortcomings were being revealed along with the government's suppression of the facts.

Yet if other editors and reporters were privately kicking themselves, most were outwardly gracious. Many promoted the work of the upstart magazine that had upstaged and outed them. Editors and reporters around the world commended Ross, Shawn, and Hersey for their courage; one *New York Times* editor even contacted Ross and called him a genius, adding, "I bow, deeply." A CBS anchor told Shawn that if an article like "Hiroshima" couldn't save the world, nothing could.

Some journalists did admit their professional jealousy to the *New Yorker* team—including one from London's *Daily Express*, which had run Wilfred Burchett's bellwether "Atomic Plague" story a year earlier. Over at the *Life* magazine offices in nearby Rockefeller Center, some of the envy was tinged with admiration; some of it was sullen. As one *Life* writer got into an elevator at the offices, carrying a copy of a white-banded *New Yorker*, another writer spotted it.

"Quite a stunt of John's, huh?" he said. "Pretty good trick. Wish I had had it."

Harold Ross was now headier than hell, as he remarked to one editor. "The story is kicking up more fuss than any other magazine story I ever heard of, and I think the excitement has only started," he told *New Yorker* writer Kay Boyle. He added to *New Yorker* writer Janet Flanner that "Hiroshima" was having a bigger success than any other magazine story published during his lifetime. To publisher Blanche Knopf he reported that he hadn't been so satisfied in years. "Hiroshima" was breaking ever bigger than he had hoped.

The magazine's employees and contributors—most of whom had learned about the "Hiroshima" issue at the same time as the rest of the world—now did newsstand sales reconnaissance work for their bosses around the city. The issue was selling out everywhere, they reported back to Ross and Shawn. One newsstand in Grand Central Terminal had posted a sign proclaiming "No More *New Yorker*s" to ward off inquirers. Another newsstand owner reported that "people rush up and say, 'You got that Nagasaki thing?' " He had saved a copy for himself; demand was so high for the issue that he thought he "could probably get a buck for this copy right now." (The magazine retailed for 15 cents.)

Within days, a low-grade black market emerged for the August 31, 1946, issue of the *New Yorker*. A friend informed Hersey that he had been trying to track down a copy in vain but finally found one at a secondhand bookstore; the magazine was offered to him for $6—and that was a bargain, he was informed.

A more solemn report also came in: one *New Yorker* contributor saw a group of Japanese American soldiers buying copies at Grand Central. The soldiers paid for the magazines, and there, amidst the bustle of rushing commuters and blaring train announcements, they sat down together and read the issue in silence.

NOW WRITE UP THE MASSACRE OF NANKING

Hersey and the *New Yorker* editors were awash in praise from their fellow journalists and editors. Now it was time to gauge the initial reaction from readers across the country. A deluge of Hiroshima correspondence, from practically every region of the United States, large cities and small towns alike, now arrived daily at the *New Yorker* offices. Editorial assistant Louis Forster had been tasked with keeping track of all of it; he noted how many people were "for" or "against" the story and reported the results regularly back to the editors.

The majority of letter writers approved of the story—which had been far from a foregone conclusion, given the widespread American support for the atomic bombings. Many of the letters indicated that Hersey and "Hiroshima" had jolted many people awake on the subject. The article was apparently changing minds fast—or at least making people very uneasy about their stance on the bomb as a necessary evil. Even George R. Caron, the tail gunner on the *Enola Gay*—the plane from which Little Boy had been dropped on Hiroshima—had called the *New Yorker* offices to ask for a copy.

Most people had a "4th of July attitude . . . in regards to the atom bomb" before Hersey's story came out, one reader wrote in. Now such celebratory bombast would be harder to defend; nor would it be easy to sanction use of the bomb again in the future. One reader wrote that he was ashamed that his tax dollars had helped commission the Hiroshima bombing. Others were having a hard time processing the fact that their country—once deemed righteous in victory—had levied this attack on a largely civilian population.

"As I read, I had to constantly remind myself that we perpetrated this monstrous tragedy," wrote one reader. "We Americans."

If some now felt pity for Hersey's six blast survivors and Hiroshima's other victims, a greater number expressed deep anxiety about the threat nuclear warfare now posed to all humans. One reader reported that he had finished the story at midnight; the rest of the night had been "fitful" and nightmare-filled. Another wrote of his fear of the "unprecedented danger of self-destruction." Hersey's plan to have his readers visualize themselves in the stead of his six protagonists was working— even if many readers were not feeling remorse for the bombings so much as fearing for their own welfare. Yet even this selfish sort of empathy could be useful if that sense of self-preservation could compel readers to action. Most correspondents stated that they thought the article was a contribution to the public good. One Pennsylvania-based woman even sent a check to the *New Yorker* office to help offset the cost of an immediate reprint of the issue. (It was returned to her, with thanks.)

However, other readers immediately canceled their subscriptions to the *New Yorker*. Some wrote in and decried the story as antipatriotic Communist propaganda, designed to undercut the United States in its moment of victory. Others called "Hiroshima" pro-Japanese propaganda. It was clearly biased, wrote one reader. It was in "very bad taste," wrote another.

"Wonderful—marvelous," read one letter. "Now write up the massacre of Nanking."

Not all editors and columnists praised the story, either. The *New York Daily News* came out swinging, calling "Hiroshima" a stunt and "propaganda aimed at persuading us to stop making atom bombs, to destroy those we've got stockpiled, and to give our technical bomb secrets away prematurely, especially to Russia." The Japanese certainly would have used it against the U.S., the newspaper contended, if that country had gotten the weapon first.

"Had we lost the race," the editorial continued, "Japanese and German reporters might now be writing tragic masterpieces about what the bomb did to a lot of San Franciscans, Chicagoans, Washingtonians, or New Yorkers."

The editor of *politics* magazine announced that he'd found "Hiroshima" so boring that he'd stopped reading it halfway through. The reader was made to feel no pity or horror for them at all, he added. Another *politics* contributor, Mary McCarthy, called Hersey's story a prime example of opportunistic catastrophe journalism, whose author had exploited "the marvelous and true-life narratives of incredible escapes."

Interview requests avalanched in from both friendly and hostile publications. As Hersey remained hundreds of miles away, sequestered in the Blue Ridge Mountains, interviewers instead descended upon Harold Ross. The *New Yorker* itself was now in the spotlight: readers were intrigued and bemused by the story of how this small, idiosyncratic humor publication—deemed too "inessential" by the government during the war to merit a higher paper quota—had gotten the blockbuster story of the war. "Notoriously the editors of *The New Yorker* are a hardboiled group," wrote one publication, which made it all the more extraordinary that they had been compelled to make magazine history with such a grave human interest war story. *Newsweek* ran a three-page behind-the-scenes "Hiroshima" story, detailing the top-secret project, from the decision to run the article in a single issue to Ross and Shawn's feeding of the text to the printers themselves.

Ross's office also received several phone calls from a young woman at *Time* magazine, Hersey's former editorial home base, requesting an interview. Ross was wary—this was, after all, his rival Henry Luce's publication, which Hersey had jilted—but he relented in the name of generating publicity for the story. The woman materialized at the *New Yorker* offices with a *Time* writer at her side. The ensuing interview was far from friendly. The *Time* duo grilled Ross unpleasantly about the workings of the office and about Hersey's time on the ground in Hiroshima.

"The two of them put on the God damnedest show I have ever seen in journalism," Ross later reported to Hersey. "The bastards were out to be mean . . . I am convinced that [the writer] was sneering most of the time." Shortly thereafter, *Time* ran its coverage of "Hiroshima," and it was everything Ross had suspected it would be.

"At 21 years of age, the New Yorker was feeling grown-up and responsible," the story began. The amateur hour *New Yorker* editors had basically stumbled into landing Hersey's scoop—which was reduced by the *Time* team to a "doomsday documentary"—and had turned it into an attention-seeking stunt. The *New Yorker* editors had only run the story, the *Time* story cynically asserted, because the magazine was in the summer doldrums. Ross was characterized as juvenile, profane, and opportunistic.

"Editor Ross, admitting to have gotten a little religion [during the 'Hiroshima' project]," concluded the *Time* write-up, "announced that he was ready to do it again if something as good came along."

If Hersey—once Luce's Time Inc. heir apparent—had thought that his seven years at that company might have earned him a more sympathetic review, he had been wrong. For Luce, Hersey was the ingrate prodigal son who had yet to make his repentant return home. The publisher was so enraged that Hersey had written the Hiroshima story for the *New Yorker* that he had Hersey's portrait removed from Time Inc.'s gallery of honor.

I COVERED IT UP

As the days passed, the sensation caused by "Hiroshima" continued to grow, propelled not just by the feverish coverage by newspapers and magazines but also by a relentless stream of radio reports. ABC Radio Network's director of public affairs, Robert Saudek, read the article and immediately contacted the *New Yorker* about airing a radio adaptation. It would be "done straight," he promised; Hersey would be given approval of the scripts. There would be no acting, no music, no effects, and no commercials—just a straightforward reading of the article, with six different actors—who would remain unidentified until the end of the final installment—reading the story of each of the six "Hiroshima" protagonists. Saudek's offer was accepted.

Joseph Julian, the actor chosen to read the parts about Reverend Tanimoto, had been dispatched soon after the bombing to Hiroshima as a Red Cross radio reporter, and had even met and interviewed Tanimoto himself long before Hersey had set foot in the city. (About being chosen to read Reverend Tanimoto's parts, Julian later stated, "I welcomed any opportunity available to render Hiroshima human, to counteract its being compressed into a paragraph of cold statistics." After seeing the bombed city, he recalled, "I knew the full meaning of the expression, 'the end of the world.'") Journalist George Hicks—who had recorded a famous D-Day radio report from the decks of the USS *Ancon*—was selected as the announcer for the series *Hiroshima*, launched on ABC at 9:30 p.m. on Monday, September 9.

"This chronicle of suffering and destruction is not presented in defense of an enemy," listeners were assured. Rather, it was being "broadcast as a warning that what happened to the people of Hiroshima, a year ago, could next happen anywhere."

The radio version of "Hiroshima" was read on-air for four consecutive nights. The telephone switchboards were swamped after the broadcasts, Saudek reported to Hersey. He told the *New Yorker* team that, to the best of his knowledge, *Hiroshima* had received the highest rating of any public interest broadcast. (When Saudek and ABC were later given a Peabody Award for the program, the Peabody committee applauded Hersey and the *New Yorker* for "their scoop of the year.") The British Broadcasting Corporation (BBC) also aired the adaptation of "Hiroshima" several weeks later, and approximately five hundred other U.S. radio stations covered the article in the days that immediately followed its release.

Many radio commentators called *Hiroshima* a cautionary tale and continued to warn that soon no one, anywhere, would be exempt from the threat of nuclear warfare. "As I read Mr. Hersey's account, it was all too easy . . . to change the Japanese names to American names," said New York–based radio commentator Bill Leonard. "All too easy to change the flimsy buildings of Hiroshima to the sturdy buildings of New York." He advised his listeners to read the article "and then maybe read it again. Because this is New York too."

One of the country's most influential commentators, Raymond Swing—whose show was carried on 135 stations nationwide—reminded his listeners that for most Americans the atomic bomb had been an abstraction and that Hersey's article brought home what these American nuclear strikes had done to fellow human beings. When America no longer enjoyed its current atomic dominance, "not a few of us in this country can expect to go through the incredible tortures that befell the figures in John Hersey's article in the *New Yorker.*" On another show, husband-and-wife radio team Ed and Pegeen Fitzgerald predicted that the article would put an end to atomic bomb humor.

"I will never joke about it again," Ed pledged.

"I will never, either," Pegeen replied.

Most listeners tuning into one particular *Hiroshima* radio show likely had no idea that one of the hosts had played a part in the government's initial Hiroshima cover-up. Lieutenant Colonel Tex McCrary of the U.S. Army Air Corps—the organizer of the first government press junket that had visited Hiroshima and Nagasaki a year earlier—had left behind the *Headliner*, the *Dateliner*, and military service. He had since moved to New York and become co-host of an NBC morning radio show, *Hi Jinx*, with his model-actress wife, Jinx Falkenburg. (Their showbiz nicknames: "Mr. Brains" and "Mrs. Beauty.")

Since his tour of post-bomb Hiroshima and Nagasaki, McCrary had privately grappled with what he had seen in those cities. The implications of the atom bomb's debut continued to trouble him. Bigger and more terrible versions of Little Boy and Fat Man would inevitably be built, he recognized, threatening civilization as he knew it. Yet, when discussing "Hiroshima" and Hersey on his radio show, he was evasive about his early role in suppressing reportage that would have helped reveal the extent of the devastation wreaked by the Japan bombs.

"You know, Tex," Falkenburg said to McCrary on their September 4 show, "when you told me the other day about the article John Hersey wrote in *The New Yorker*, and when I heard that every copy sold off the newsstands a day after publication, I couldn't understand it. Why all this sudden interest in something that happened a year ago? Hiroshima is an old story now."

"In one sense, yes, Jinx," McCrary replied. "[But] when you realize that atomic power is not yet conquered or in chains, you know that Hiroshima is not an old story. It was never newer than now . . . the story of Hiroshima and Nagasaki, the story of the once top-secret Manhattan Project, is still as newsworthy as destiny itself."

He gave an abbreviated account of his own experience of seeing Hiroshima with those newsmen he had hastily escorted through the ruins: they were hardened American war correspondents, he stated, but they had all been shocked by what they'd seen. It had been a terrible day for all of them.

However, McCrary did not mention the "censored" stamp that had hung in his junket planes or his role in tempering the whole truth about the nuclear aftermath in Hiroshima. McCrary also did not discuss the near-total suppression of reporters in General MacArthur's Japan or the threat of expulsion or imprisonment those reporters had faced for bomb reporting that displeased the government. Rather, McCrary explained to his listeners, Hersey had gotten the Hiroshima scoop instead of these early-arriving junket journalists because Hersey was simply "more than a reporter" and had managed to tell it better.

Later, though, McCrary did finally admit that he had obstructed the junket reporters from writing fully about what they had seen in the bombed city.

"I covered it up, and John Hersey uncovered it," he stated. "That's the difference between a P.R. man and a reporter."

AN IMMEDIATE PLACE IN HISTORY

During the war, Hollywood had gone into overdrive creating films that portrayed the Japanese as the "yellow peril." Now, many of its executives tried to capitalize on the success of "Hiroshima." Upon the article's release, the *New York Daily News* sourly reported that movie executives "had come swarming around [Hersey], their fists congested with offers." Fists congested or not, the inquiring Hollywood execs were rebuffed. Hersey—along with the rest of his team at the *New Yorker*—had decided that there were to be no radio dramatizations, and that "for the time being, disposition of movie and dramatic rights is not subject to discussion." Producers, agents, and studio executives had indeed conveyed their admiration and been asking after the film rights; within a couple of weeks Hersey had been approached with several offers.

Hersey was quickly becoming as well-known as the actors who might have been tapped to star in a major film production of "Hiroshima." If he had been famous before "Hiroshima" as a Pulitzer Prize–winning author, his fame had now jolted to another level. Many publications ran profiles and photos of Hersey alongside their "Hiroshima" coverage. By late fall, he was notified that he had been designated by the Celebrity Information and Research Service, Inc., as one of the "Ten Outstanding Celebrities of 1946," along with Army chief of staff and former U.S. supreme commander General Dwight D. Eisenhower, singer Bing Crosby, and actors Laurence Olivier, Joan Crawford, and Ingrid Bergman. (When Hollywood gossip columnist Louella Parsons did a radio broadcast announcing the lineup, Ross snickered to Hersey and Shawn that "Miss Parsons did not attempt to pronounce 'Hiroshima.'")

Some of the *New Yorker* readers who sent cables and letters to the magazine called for Hersey to win another Pulitzer Prize for his Hiroshima reporting. Atomic Bill Laurence of the *New York Times* had just won a Pulitzer Prize in reporting, for his government-sanctioned "eye-witness account of the atom-bombing of Nagasaki and his subsequent ten articles on the development, production, and significance of the atomic bomb." Ross had disappointing news for fans who hoped that "Hiroshima" would be similarly honored: Hersey wasn't eligible, he informed them, because the reporting Pulitzer was only awarded to newspaper stories.

The Library of Congress immediately made a bid to get the first draft of "Hiroshima." Upon the article's release, the director of acquisitions told Hersey that it was one of the notable documents of modern times. Even if Hersey had declined to promote the article upon its release, he was apparently legacy-minded enough to pledge the first draft of "Hiroshima"—along with his notes—to the library of his alma mater, Yale University, whose representatives were so excited that they issued a press release announcing the acquisition. (The gift initially rankled Harold Ross when he found out about it. "How did [Hersey] get this [draft], may I ask?" he quizzed an officer manager. It's unclear if he got his answer.)

"Hiroshima" had already created more of a stir than any article before it, and its posterity was about to be secured in book form. Even before the August 31 issue had been finalized, Ross had sent proofs of the article to Alfred A. Knopf, Inc., which had published Hersey's books *A Bell for Adano*, *Men on Bataan*, and *Into the Valley*. "That book ought to sell like hell," Ross told Alfred Knopf, who prepared to rush out to the "widest market possible" a first printing of 50,000 copies to be released on November 1. Another deal had also been immediately set in place with the Book-of-the-Month Club, which was also working fast to release it simultaneously with the Knopf edition. The Book-of-the-Month Club gave the book a huge platform, sending out an announcement about *Hiroshima* to nearly a million of its members. It was, the club asserted, "destined to be the most widely read book of our generation."

"It is hard to conceive of anything being written that could be of more importance to the human race," it added.

In England, Penguin Books also prepared to publish a quarter of a million copies of its own edition of *Hiroshima*. When released, it sold out within weeks. Hersey's story was becoming a global phenomenon: it had been serialized in newspapers around the world and—as darkly noted by Knopf, and much to the *New Yorker*'s annoyance—it had been pirated by newspapers around the world. A China-based editor, Randall Gould, wrote to Hersey to tell him that he had seen the story appear in the *Shanghai Evening Post*. It would have been impossible to arrange official permissions for the reprint there, Gould said, "because as you know there is no copyright here. China likes to steal things." Regardless, he offered his congratulations.

Hiroshima, Gould told Hersey, had "put you in History."

Chapter Seven

Aftermath

AN IMAGE PROBLEM

General MacArthur's SCAP offices had approved Hersey's entrance into Japan and given him access to Hiroshima. FBI officials in both Tokyo and Washington had known of his presence in the country. Hersey had stayed with U.S. military police while in Hiroshima. General Groves had complete pre-publication knowledge of the contents of the article. Even so, "Hiroshima" blindsided officials at the very highest levels of government. They quickly and painfully discovered, along with the press, that the bombings of Hiroshima and Nagasaki were not yesterday's news; nor had their efforts at spinning and supressing that news succeeded after all.

"We all exhausted ourselves" reading the article, McGeorge Bundy, former assistant to U.S. secretary of war Henry L. Stimson, admitted later.

"Hiroshima" undid over a year of work on two continents to cover up the truth about the aftermath of the bombings of Hiroshima and Nagasaki. General Groves's right-hand man, General Thomas F. Farrell—who had first inspected those cities for residual radiation a year earlier and declared both safe for incoming occupation troops—was incensed by Hersey's story, apparently unaware that his own wartime boss had greenlighted it.

"America forgets so quickly," General Farrell wrote in a letter to Bernard Baruch, the U.S. representative to the United Nations Atomic Energy Commission. Farrell said he was "much more moved by starved American soldiers who had been continually beaten by baseball bats than I was by the wounded Japanese in Hiroshima."

Baruch personally knew Ross and Shawn and *New Yorker* publisher Raoul Fleischmann; General Farrell now urged him to tell the magazine's principals to run a similar article on six Allied prisoners of war. These POWs should be able to describe their brutal treatment by their Japanese captors and be allowed to give *their* thoughts on the use of the atomic bombs.

General Groves made no public statement about his unlikely role in bringing "Hiroshima" to the masses. He did, however, find immediate practical applications for Hersey's article. Not long after the issue came out, William Shawn received a letter from a War Department public relations officer informing him that General Groves had just mentioned Hersey and "Hiroshima" in a speech to the Command and General Staff School at Fort Leavenworth, Kansas. Among the general's topics that day: the future of nuclear warfare and the United States' need for preparedness for possible atomic war with new enemies. The Army needed to ready itself for a new role if the United States were ever attacked with nuclear weapons; American ground forces now needed to study the nuclear attacks on Japan to learn how to "aid and control the populations of our own atomically bombed cities."

"The catastrophe of an atomic attack, I am afraid, has never fully been brought home to us," General Groves stated.

To that end, he declared, all present ought to read John Hersey's "Hiroshima." In fact, the article should, in his opinion, be required reading for all American officers, for its depiction of nuclear aftermath would be an invaluable tool in helping to prepare a highly trained and equipped military response to future attacks. (The press officer who flagged the speech to Shawn also informed the editor that there was enormous demand for the story within the military.)

On the other side of the Pacific, General MacArthur also found comparable uses for the article. His immediate reaction upon reading "Hiroshima" is unknown; it was surely lost on no one in Washington, D.C., that SCAP's GHQ had given Hersey access to Hiroshima in the first place. Embarrassingly for the U.S. government, to the rest of the world—for whom "Hiroshima" had all of the qualities of a total exposé—it likely looked as though Hersey had sneaked into the country and gotten the story undetected, right under SCAP's nose.

Yet like his competitor General Groves, General MacArthur himself evinced no public embarrassment or indignation about the story. In due course, the *New Yorker* was contacted by another War Department public relations officer with a request: Could Hersey be persuaded to release rights to "Hiroshima" for a special edition of the article?

"It is my understanding," the officer wrote, "that General MacArthur plans to reproduce the piece to be used for instructional purposes within the Army in the Far East Theater."

Despite the instructional military utility General MacArthur and General Groves were finding in Hersey's story, on the whole the U.S. government—victor over fascism and tyranny—suddenly had a serious post-"Hiroshima" image problem. The transition from global savior to genocidal superpower was an unwelcome reversal. Hersey's readers around the world were now reevaluating America's superior moral stature and demanding to know why these revelations had taken more than a year to come out. If something of this enormity had been successfully hidden from the public, what else was being hidden? What other information about these new weapons was the U.S. government concealing, and had Hersey been accurate when he wrote in "Hiroshima" that more powerful, terrifying versions of it were being developed?

The livid "Hiroshima"-inspired editorials kept coming. A couple of weeks after the story's release, *Saturday Review of Literature* columnist Norman Cousins wrote a scathing column in response to Hersey's story that particularly incensed senior members of the government.

"Do we know . . . that many thousands of human beings in Japan will die of cancer during the next few years because of radioactivity released by the bomb?" he wrote. In reality, the atomic bomb was a death ray and its use on humans a crime, he stated. Furthermore, the United States had all but guaranteed use of atomic weapons in the next war—without considering that America's own population was especially vulnerable to extreme damage, given its dense population centers. Hersey and the *New Yorker* had provided a crucial wake-up call, but Americans were facing a full-blown crisis, Cousins wrote, and everyone in the country needed to acknowledge and address the Pandora's box that the U.S. had opened that previous summer.

The avalanche of damaging criticism started to pile in from important figures outside the press as well. Fleet Admiral William F. "Bull" Halsey Jr., commander of the Third Fleet in the Pacific, stated in a news conference that the dropping of the bomb had been an unnecessary experiment and a military mistake.

"Why reveal a weapon like that to the world when it wasn't necessary?" he said. "[The U.S.] had this toy and they wanted to try it out, so they dropped it. It killed a lot of Japs, but the Japs had put out a lot of peace feelers through Russia long before."

Some of the scientists who had built "this toy" had also been publicly confessing their misgivings about their role in creating it. Before the bombs had even been dropped on Japan, a group of senior Manhattan Project scientists had privately lobbied against deploying them and begged the U.S. government instead to make a demonstration of the bomb's might. In using the atomic bomb to attack Japan, the United States "would sacrifice public support throughout the world, [and] precipitate the race of armaments," they warned. Not long after the bombs had been dropped on Hiroshima and Nagasaki, the Manhattan Project's J. Robert Oppenheimer—dubbed "father of the atomic bomb"—announced his own feelings of conflict.

"If atomic bombs are to be added as new weapons to the arsenals of a warring world, or to the arsenals of nations preparing for war," he said in a speech, "then the time will come when mankind will curse the names of Los Alamos and of Hiroshima."

Physicist Albert Einstein—whose formula, $E=mc^2$, gave researchers a way to quantify the vast potential energy that would be released in a nuclear explosion—had long been trying to sound the alarm about the dangers posed by nuclear weapons. Just before the war he had alerted President Franklin D. Roosevelt to Germany's alarming efforts to create an atomic bomb. Einstein had been uninvolved in the Manhattan Project and disavowed any personal fatherhood of the bomb; the creation of nuclear weapons posited, in his opinion, a catastrophic threat to the survival of humanity. He had fearfully predicted that, after the war, countries across the globe would race to become nuclear powers, which could lead to even more horrific destruction.

"Today, the physicists who participated in forging the most formidable and dangerous weapon of all times are harassed by an equal feeling of responsibility, not to say guilt," Einstein had said in a post-Hiroshima and -Nagasaki speech at New York's Hotel Astor, just around the corner from the *New Yorker* offices. Just weeks before "Hiroshima" hit newsstands, Einstein had told the *New York Times* that rockets could now carry atomic bombs, making virtually every center of population on earth vulnerable to devastating nuclear attack. He called on Americans to immediately halt their everyday activities to consider the implications of Hiroshima and the start of the atomic age.

"To the village square we must carry the facts of atomic energy," he said. "From there must come America's voice . . . We cannot leave it to generals, Senators, and diplomats."

Einstein now contacted the *New Yorker*'s publisher, Raoul Fleischmann, about "Hiroshima." He expressed his deep admiration for the work and requested a thousand reprints of the story to distribute to leading scientists around the world. His request was fulfilled.

"Mr. Hersey has given a true picture of the appalling effect on human beings . . . subjected to the unprecedented destruction achieved by the explosion in their midst of one atomic bomb," Einstein wrote in the cover letter he sent out with his copies of "Hiroshima." He added, "This picture has implications for the future of mankind which must deeply concern all responsible men and women."

For certain U.S. officials and unrepentant Manhattan Project principals, these waves of criticism were as damaging as those initial press reports out of Japan in the earliest days of the occupation. Immediate action needed to be taken to contain and spin the story—again.

STRAIGHTENING OUT THE RECORD

Harvard president and Manhattan Project advisor James B. Conant had just returned from a nearly monthlong holiday in New Hampshire's White Mountains when he read Hersey's story. "Hiroshima" alarmed him greatly. The article was turning public opinion against the bomb and its creators—not to mention that it was undermining Americans' belief in their leadership. It had not only exposed how much the government had withheld from the U.S. public and eroded the country's moral standing, it could—contrary to General Groves's hopes—jeopardize public support for future nuclear arsenal building and preparedness.

Conant, a chemist, had a deep history of advancing the science of warfare. During World War I, he had helped spearhead production of

John Hersey at his desk at *Time* magazine in early 1945. Then just thirty years old, he already had an enviable career as an international correspondent throughout the war. He would go on to win the Pulitzer Prize later that year for his 1944 novel, *A Bell for Adano*.
Photo by Time Life Pictures/Pix Inc./The LIFE Picture Collection via Getty Images.

The mushroom cloud following the detonation of Little Boy—a nearly 10,000-pound uranium bomb—over Hiroshima. The cloud would reach tens of thousands of feet into the sky. *Photo by Time Life Pictures/US Army Air Force/The LIFE Picture Collection via Getty Images.*

The ruins of Hiroshima immediately after the atomic bombing. The estimated death toll has ranged from 78,000 to 280,000, although the exact number of dead and injured will never be known. *Photo by Hulton-Deutsch/Hulton-Deutsch Collection/Corbis via Getty Images.*

August 15, 1945: Two million people crowded into New York City's Times Square to celebrate Japan's defeat. When the *New York Times* announced the official surrender on its electric zipper sign there, "the victory roar . . . beat upon the eardrums until it numbed

the senses," recalled one *Times* correspondent. The party was "instantaneous and wild," and the "metropolis exploded its emotions with atomic force." *Photo by © CORBIS/Corbis via Getty Images.*

U.S. General Douglas MacArthur—now Supreme Commander for the Allied Powers—arriving at the Atsugi Airdrome at the beginning of the occupation of Japan on August 30, 1945. To many in Japan, it seemed that the country now had two emperors. *Photo by Time Life Pictures/US Army/The LIFE Picture Collection via Getty Images.*

In Tokyo, General MacArthur established his general headquarters in the fortress-like Dai-ichi Life Insurance Company building right across the street from the Imperial Palace, where the Japanese emperor still dwelled. It was not a subtle statement on MacArthur's part. *Photo by Paul Popper/Popperfoto via Getty Images/Getty Images.*

United Press correspondent Leslie Nakashima was the first journalist affiliated with a western news outlet to get into Hiroshima after the bombing, on August 22, 1945. In a wire story, he reported that Hiroshima was a wasteland of ash and rubble. The *New York Times* ran only a heavily edited version of his account. *Used with permission from the Nakashima/Tokita family.*

Veteran war correspondent Wilfred Burchett landed in Japan with an early wave of Allied occupation forces and immediately made his way to Hiroshima, even though Western correspondents had been forbidden by occupation authorities to travel throughout the country. His subsequent *Daily Express* report, "Atomic Plague," depicted the radiation effects killing blast survivors and sent ripples of alarm around the world. A U.S. official accused him of having "fallen victim to Japanese propaganda." *With permission from the Wilfred Burchett Estate.*

Chicago Daily News correspondent George Weller (*left*) also got into Japan with early-arriving occupation forces and separately made his way into Nagasaki, which had been devastated by a nuclear bomb on August 9, 1945—three days after the Hiroshima bombing. Weller's report on the destruction there was intercepted and "lost." Occupation authorities quickly declared the atomic cities off-limits to foreign reporters. *Used with permission from Anthony Weller.*

Lieutenant Colonel John Reagan "Tex" McCrary, a reporter turned public relations officer for the U.S. Army Air Forces, took a junket of hand-picked Allied reporters into Hiroshima, but advised the journalists to downplay what they had seen there. "I don't think they're ready for it back home," he told them. *Used with permission from Michael McCrary.*

Manhattan Project leaders Major General Leslie R. Groves and physicist J. Robert Oppenheimer at the New Mexico site of the July 16, 1945, atomic bomb test detonation. On September 9, the men led a press tour of the site to downplay the bomb's aftereffects. "The Japanese claim that people died from radiations," General Groves told reporters. "If this is true, the number was very small." *Photo by Rolls Press/Popperfoto via Getty Images/Getty Images.*

During late summer and fall 1945 visits to Manhattan Project labs and industrial contractors, General Groves—shown here at the atomic research facility in Oak Ridge, Tennessee—gave speeches in which he told his audiences that there was no need for guilty consciences over the atomic bombings. He personally had no qualms, he said. *Photo by PhotoQuest/Getty Images.*

William Shawn, the *New Yorker*'s deputy editor and the magazine's "hunch man," was shy and introverted, but also charismatic. To him, "every human being [was] as valuable as every other human being, . . . every life was sacred," recalled one of his former writers. *Photograph by Lillian Ross. Used with permission of the Lillian Ross Estate.*

John Hersey on assignment in China in 1946, right before flying to Japan in May to report on Hiroshima. *Photo by Dmitri Kessel/The LIFE Picture Collection via Getty Images.*

New Yorker cofounder and editor Harold Ross was an extravagant personality with a talent for profanity—and an expert at getting exclusives, even out of journalist-flooded war-devastated landscapes. "How simple it is to scoop the world," he told one of his reporters during the war. *Photo by Bachrach/Getty Images.*

The *New Yorker*'s offices—spread out over a few floors of 25 West 43rd Street—were ramshackle by anyone's standards. Standing in the elevator lobby was a brazier stuffed with cigarette butts and wadded-up rejection slips. *Taken by Hobart Weekes; used with permission from James McKernon.*

Harold Ross's office at the *New Yorker*. For ten days, Hersey, Ross, and Shawn sequestered themselves in the locked room to edit "Hiroshima" in secret. The undertaking was their own journalistic version of the Manhattan Project. *Taken by Hobart Weekes; used with permission from James McKernon.*

Father Wilhelm Kleinsorge, the first of Hersey's chosen six protagonists in "Hiroshima." A Hiroshima-based German priest, Kleinsorge acted as a translator for Hersey and introduced him to other Hiroshima *hibakusha*, or atomic blast survivors. *Asahigraph, 1952.*

Reverend Kiyoshi Tanimoto of Hiroshima, the second of Hersey's protagonists and a "rescuing angel" who courageously evacuated bombing victims to safer ground as fires engulfed the city. *Asahigraph, 1952.*

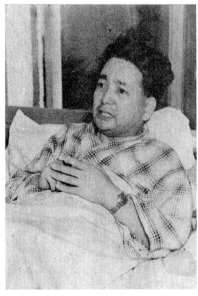

Hersey's third "Hiroshima" protagonist, Dr. Terufumi Sasaki, had been on call at his Hiroshima hospital on the day of the bombing. One of the few Hiroshima medics to survive unscathed, he treated hundreds of patients in the days that followed. By the third day, most of the patients he had treated earlier had died. *Asahigraph, 1952.*

Dr. Masakazu Fujii, Hersey's fourth protagonist, had run a small private hospital, before it collapsed on him during the bombing. *Asahigraph, 1952.*

Hersey's fifth protagonist, a young widow named Hatsuyo Nakamura, had been preparing breakfast in her home when the bomb exploded, burying her three children in debris. *Asahigraph, 1952.*

Miss Toshiko Sasaki, a young clerk and Hersey's final protagonist, was nearly crushed to death by felled office bookshelves when the bomb went off. *Asahigraph, 1952.*

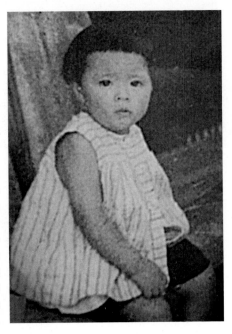

Koko Tanimoto, the daughter of Reverend Tanimoto and his wife, Chisa, around the time of the bombing. When the bomb exploded, Chisa was inside the family's home, holding Koko; the entire structure collapsed on top of them. Although Chisa's arms had been pinned to her sides by debris, she managed to free one and scratch out a hole through the rubble. Soon it was just big enough to push Koko through. Both survived. *Used with permission from Koko Tanimoto Kondo.*

The August 31, 1946, issue of the *New Yorker* containing "Hiroshima" and almost nothing else. All of the other usual features had been cut. The incongruous cover scene carried unnerving connotations for readers: It depicted a sleepwalking America that had indeed "escaped into easy comforts," as Albert Einstein had put it just months earlier, while ignoring the perils of the atomic age. *Used with permission from the* New Yorker *magazine.*

President Harry Truman being briefed on the Hiroshima bombing by Secretary of War Henry L. Stimson. After Hersey's "Hiroshima" was released, Truman did not give a public statement about the story, but tasked Stimson with "straightening out the record" amidst the controversy stirred up that fall. Stimson duly published a retort story in *Harper's Magazine*, anonymously coauthored by many figures from the "old-boy War Department network." *Bettmann / Contributor*

In the year since leading his post-bomb Hiroshima press junket, Tex McCrary had become co-host of an NBC morning radio show, *Hi Jinx*, with his model-actress wife, Jinx Falkenburg. When Hersey's "Hiroshima" came out, McCrary devoted two show segments to the seismic story, but did not reveal his previous role in covering it up. Later, however, McCrary would confess that "I covered it up, and John Hersey uncovered it." *Photo by Leonard McCombe/The LIFE Images Collection via Getty Images/Getty Images*

poison gas. During World War II, he had been tapped by President Roosevelt as a lead scientist to help helm the Manhattan Project. (Some had dubbed him a "Grand Duke" of the Project). Now, in the autumn of 1946, he was growing disgusted by his fellow scientists professing guilt about their part in creating the bomb. ("I wept as I read John Hersey's *New Yorker* account," admitted one Manhattan Project scientist. The story had filled him with shame when he recalled the "whoopee spirit" with which the Manhattan Project principals had reacted to news of the Hiroshima bombing a year earlier.) Conant, on the other hand, was unrepentant about urging the use of nuclear weapons on Japan.

"War is ethically totally different from peace," he later stated.

Conant quickly wrote a letter to Harvey H. Bundy (McGeorge Bundy's father), who had been special assistant on atomic matters to Secretary of War Henry Stimson. There had been a lot of dangerous "Monday morning quarterbacking" about the Japan bombings lately, Conant wrote. To that end, he enclosed a clip of the incensed *Saturday Review of Literature* "Hiroshima"-inspired editorial by Norman Cousins. All of this concern about the Japan bombings was just sentimentality, Conant told Bundy, and even though he was convinced that "all this talk" was only a minority position, it was, unfortunately, a vocal minority.

"It seems to me of great importance," Conant maintained, "to have a statement of fact . . . clarifying what actually happened with regard to the decision to use the bomb against the Japanese."

If something wasn't done to counter the bad publicity being put out by this "group of so-called intellectuals"—i.e., writers like Hersey and Cousins—they might influence the opinions and impressions of the next generation, and distort history.

All of the bad Hiroshima press was also disturbing President Harry Truman. The president had grown especially irritated by allegations that the United States had bombed Hiroshima and Nagasaki recklessly and without elaborate consideration.

"The Japanese were given fair warning, and were offered the terms which they finally accepted, well in advance of the dropping of the bomb," he wrote to physicist Karl T. Compton, a member of the government's Interim Committee—a committee that had been formed to advise President Truman on military use of nuclear weapons, and which had recommended the atomic attacks on Japan. "I imagine the bomb caused them to accept the terms," Truman added.

Yet the White House remained publicly silent on the subject of Hersey's "Hiroshima," perhaps in an effort to downplay its significance, and much to Harold Ross's annoyance. Then the *New Yorker* editor read an item in the *New York Post* stating that President Truman had been quizzed about whether he'd read the Hersey story that was creating such a furor.

"I never read *The New Yorker*," the president was quoted as saying. "Just makes me mad."

Ross decided to kick the hornet's nest. He sent a letter—and eventually three copies of the "Hiroshima" issue—to the president's press secretary, Charles G. Ross, and urged him to call President Truman's attention to the story. Charles Ross replied cordially enough—he personally was a longtime fan of the *New Yorker*, he stated—but was coy on the question of whether the president had read or was aware of Hersey's story.

"The President may or may not have read the Hersey article," Charles Ross wrote. "I will see that he gets the magazines you have sent me."

President Truman, like Conant, had also come to the conclusion that the record on the atomic bombs had to be quickly set straight again. The national—and international—conversation needed to be steered away from Hersey's images of the charred human remains and young families suffering from radiation sickness. The government's official talking points had to be reasserted: The atomic bombs dropped on Japan had shortened the war. The Japan atomic bombs had saved lives on both sides. Japan would never have otherwise surrendered without a protracted, bloody fight to the last man.

President Truman quietly approached former secretary of war Henry L. Stimson—who had retired just weeks after the Japan-U.S. surrender ceremony in September 1945—about pulling together some sort of public statement. "I have asked . . . Stimson to assemble the facts and get them into record form," he wrote to Karl Compton.

James Conant had also concluded that there was "no one who could do this better than Mr. Stimson," and had approached the former war secretary separately. Stimson was then working on his memoirs on his Long Island estate. Conant went to Stimson's house and, over lunch, campaigned for the former statesman to undertake the reset statement—which, Conant felt, should be issued in article form like "Hiroshima." It could not look like a patent attempt to direct attention away from the atomic bombs' cruel aftereffects nor a scrambling bid to reclaim the postwar moral high ground. It had to exude calm authority, courtesy of a trusted figure who was magnanimously, even indulgently, quelling a bit of unnecessary hysteria.

"The statement should be largely factual and not an attempt to argue too much as to the military necessity for the bomb," Conant advised.

Stimson was a savvy choice to commandeer the Hiroshima conversation and serve as a national balm. Since 1943 he had served as the president's senior advisor on the military employment of atomic energy, and had helped select Hiroshima as a target for the bomb. He projected the image of calm reassurance and was described once by the *New York Times* as displaying "an integrity that to his friends sometimes bordered on the painful."

Stimson agreed to undertake the task, but with deep private reservations. He had been selected as the "victim" by Conant to combat all of the negative press, he told a friend. Another friend of his later recalled that Stimson had "[lain] awake at night before the decision [to drop the bomb] thinking about the consequences of dropping it on a civilian target, a city of that size." Stimson had already fretted over the earlier firebombing of Tokyo: "I did not want to have the United States get the reputation for outdoing Hitler in atrocities," he admitted. Conant's request prompted him to some anguished soul-searching.

Once the project was set in motion, "the old-boy War Department network went to work" on drafting the rebuttal, as one Conant biographer put it. Input and language was solicited from Harvey Bundy, former Stimson aide George L. Harrison, and War Department historian Rudolph A. Winnacker, among others. Harvey Bundy's son McGeorge served as Stimson's aide on the project.

On November 6, General Leslie Groves sailed in with comments for the retort article. The statement was, he wrote, "a splendid description of the entire project in very condensed form." Such an article was necessary now more than ever, he told his former colleagues, and made efforts to emphasize his own outsized role in the whole matter of the bomb-dropping decision. It is unclear whether General Groves revealed to the others on the Stimson statement team that he had personally cleared Hersey's article for publication.

It was suggested that the team approach Henry Luce's *Life* magazine about running the Stimson article. This publication had an enormous circulation, but Conant proposed *Harper's* magazine, which had more gravitas. *Harper's* had also given serious attention to "Hiroshima," calling it a "staggering report," and had even defended Hersey against the few negative reviews it received. When approached by the Stimson team, however, the magazine agreed to feature the Stimson article—now titled "The Decision to Use the Atomic Bomb"—as its cover story for February 1947.

Up through the final edits, Conant appeared to be consulting the John Hersey editorial playbook, instructing his coauthors to steer away from emotional entreaties and hyperbolic language and to stick to an apparent "mere recital of facts." They should "eliminate all sections in which the secretary appears to be arguing his case or justifying his decision," he directed. "It will be very hard for anyone on the other side to challenge this article if [it] deals almost entirely with the facts."

When Conant saw the final article in an advance copy of *Harper's*, he was pleased with the team's handiwork. It was exactly right, he told Stimson, adding that they could not have allowed "the propaganda against the use of the atomic bomb . . . to grow unchecked."

Regardless, Stimson had an eleventh-hour attack of nerves as the article was being finalized.

"I have rarely been connected with a paper about which I have so much doubt at the last moment," Stimson told a friend. "I think the full enumeration of the steps in the tragedy will excite horror among friends who heretofore thought me a kindly-minded Christian gentleman but who will, after reading this, feel that I am cold blooded and cruel."

A subsequent letter from President Truman may have soothed the appointed soother, or at least steeled him. "I think you know the facts of the whole situation better than anybody," he told Stimson, reminding him that he was being charged with "straighten[ing] out the record on it." Stimson replied that he hoped the story would counter that "rather difficult class of the community" that had been making so much trouble for them.

Thus fortified, Stimson personally sent a copy of the draft to Hersey's rejected mentor, Henry Luce, who could likely be relied upon to promote it with particular zeal.

WE DESERVE SOME SORT OF MEDAL

If Conant and Stimson and their team had followed the *New Yorker*'s example with its deadpan, just-the-facts approach in their article, the *Harper's* editors took the opposite strategy from Harold Ross and William

Shawn in creating its Stimson issue cover. The Stimson article was teased in large black letters against an arresting red-and-white background:

Henry L. Stimson

FORMER SECRETARY OF WAR

Explains Why We Used

The Atomic Bomb

As with Hersey's article, "The Decision to Use the Atomic Bomb" by Stimson and company created its own media sensation. If "Hiroshima" had shredded American consciences raw, the Stimson article was a welcome, albeit narcotic, salve. Many Americans had craved official reassurance after Hersey had turned the spotlight on the fact that their government and military had perpetrated a nuclear holocaust and tried to whitewash the results.

The Stimson story in *Harper's* delivered new and improved variations of the government's official themes, laced with seeming admissions and revelations of its own. Yes, the atomic bombs levied in their name had been devastating, Stimson conceded. But they had been absolutely necessary—"our least abhorrent choice" when it came to ending the war with Japan. In Stimson's retelling, the nuclear option was once again depicted as humane. Unlike General Groves, he didn't attempt to convince Americans that radiation poisoning was a "very pleasant way to die," but he maintained that the bomb had saved the Japanese from themselves. He and other decision makers had reasoned that the extreme shock delivered by the bombs seemed the surest way to compel the Japanese to surrender, and in the process "save[d] many times the number of lives, both American and Japanese, that it . . . cost." It had also spared the Japanese further fire raids and relieved them of the "strangling blockade" imposed on the islands by Allied forces.

In a reductive counterplay, Stimson did not mention Hersey by name, nor any of those who had written outraged or anguished editorials in response to "Hiroshima." Rather, he made it sound as though the national outcry had been more of a muted objection. "In recent months there has been much comment about the decision to use atomic bombs" on Japan, he stated in a garden party tone, adding that his own comments were intended to address the concerns of "all who may be interested."

One of the highlights of Hersey's "Hiroshima" had been the statistic of 100,000 Japanese killed by the bomb. Stimson now offered up a few newsy counter-statistics of his own. In July 1945, Stimson stated, U.S. intelligence had projected that the Japanese Army was still 5,000,000 strong and had available 5,000 suicide aircraft. He reminded readers that Japan's military men belonged "to a race which had already amply demonstrated its ability to fight literally to the death." He had been informed, he said, that a land invasion "might be expected to cost over a million casualties, to American forces alone." (Even though President Truman had been provided with July 1945 military documents predicting that a full-fledged invasion might result in 40,000 American military deaths and 150,000 wounded, the 1 million casualties number was reiterated throughout the piece.) Stimson did not address Admiral Halsey's public statement that the "Japs had put out a lot of peace feelers" and had already been trying to surrender via entreaties made to the Soviets, who at the time had been official allies to the Americans.)

Why hadn't the United States simply given a demonstration of the bomb in some uninhabited area to compel surrender? The idea had been "discarded as impractical," Stimson said. Because its makers were not yet fully familiar with the weapon they had created, he wrote, they had not even known whether it would detonate when dropped from a plane, and "nothing would have been more damaging to our effort to obtain surrender than a warning or a demonstration followed by a dud."

And anyway, the U.S. hadn't had a lot of bombs to spare for such demonstrations, Stimson explained. The Japanese needed to think there was an unlimited supply, and that after Nagasaki the United States could continue to drop one atomic bomb after another until the country was obliterated. The "dread of many more" incoming bombs was what demoralized the Japanese into capitulating. In reality, in August 1945, the U.S. had only had two workable bombs, Stimson stated, which they dispensed on Hiroshima and Nagasaki—but luckily the bluff had worked. In this sense, the atom bomb not only proved to be "a weapon of terrible destruction" but also a "psychological weapon."

Ultimately, the U.S. decision had been the right one, he asserted: "All of the evidence I have seen indicates that the controlling factor in the final Japanese decision to accept our terms of surrender was the atomic bomb." He did not mention that even the publicly available part of the government's own U.S. Strategic Bombing Survey stated that "it cannot be said . . . that the atomic bomb convinced the leaders who effected the peace of the necessity of surrender" and added that the decision to surrender had actually been put into place in May—three months before the bombs were dropped.

The Stimson article essentially sidestepped the agonies inflicted on victims in Hiroshima and Nagasaki, and did not even acknowledge the radioactive qualities of the atomic bombs other than stating that they had "a revolutionary character" and a "generally unfamiliar nature." Stimson wrote that, in his mind and in the minds of those who had helped build it, the atom bomb had been "as legitimate as any other of the deadly explosive weapons of modern war." In fact, it had been seen as such a casualty preventer that "no man in our position and subject to our responsibilities, holding in his hands a weapon of such possibilities for accomplishing this purpose and saving those lives, could have failed to use it and afterwards looked his countrymen in the face."

As the Stimson story was considered the first official account of the deliberations behind the use of the bombs, hundreds of the news outlets that had covered "Hiroshima" now raced to cover the Stimson story. President Truman commended Stimson on his job, telling him that he had done "very well" with the record straightening. The authors of the story also congratulated Stimson—and themselves.

"We deserve some sort of medal," McGeorge Bundy wrote to Stimson, "for reducing these particular chatters to silence."

THE INERADICABLE FACTS

The Stimson article did not silence all of the "chatters," however. Nor did it snuff out protests, erase images of Hiroshima's obliteration, or quell widespread anxiety and horror over the prospect of future nuclear conflict. On the contrary, the influence of "Hiroshima" continued to spread after the retort in *Harper's* came out. Many readers did find relief and reason in the *Harper's* explanations. But for others the Stimson statement did nothing to address the urgent questions that had been raised during the lead-up to its release. Nor did it address the depth of the government's cover-up about the bombs and their effects.

Publications that a year earlier likely could have been relied upon to accept and promote the Stimson report were now more guarded in their responses. This new caution indicated the beginning of a rift between the wartime partners: the press might have acted as a team player with the government and military during the war, but it was now growing more judgmental of those teammates a year later.

Henry Luce's *Time* and *Life* declined to give significant coverage to the Stimson story after all, beyond an item in *Time*'s "National Affairs" section summarizing the main points Stimson had made in his story. The *New York Times* afforded Atomic Bill Laurence his usual front-page space to cover this latest development in the nuclear saga, but he did so with uncharacteristic restraint and without pseudo-biblical commentary. The *Times* editorial board itself grappled with the assertions made in the Stimson account. The newspaper's editors accepted the Stimson article's assertions that the atomic bomb had not been created or used on a whim, and agreed with the former war secretary that the Hiroshima and Naga-saki bombs had caused the Japanese to surrender.

Yet the *Times* noted that there were sure to be more questions and objections about whether the government had indeed given the Japanese fair warning, as President Truman maintained, having never given a demonstration of the bomb before using it on Hiroshima, and chided Stimson for justifying the bombs' use based on the practical results alone. Doing so was no different from "the German military thesis that necessity knows no law," the editorial read, or "that the most brutal war is the most merciful because it ends more quickly, and that the end justifies the means."

For the government, the *New York Times* had once again offered up a disturbing bellwether. The newspaper had not only indicated a new editorial aloofness; it also made clear that the Stimson retort had failed to address the painful revelations of Hersey's "Hiroshima"—namely, the horrific human cost exacted by the weapon and the fact that the aftermath reality had been covered up by its perpetrators. The United States had used an experimental radioactive mega-weapon that was still causing its civilian victims to die months after its detonation. This was now, thanks to Hersey and his team, an undisputed, ineradicable fact. Stimson had conceded in his article that the Hiroshima bomb had "brought death to over a hundred thousand Japanese." He wrote that he did not wish to "gloss over it," but had done precisely that.

"The face of war is the face of death," Stimson stated, "[and] death is an inevitable part of every order that a wartime leader gives."

Part of the Stimson article gloss was its effort to return the story of Hiroshima to the realm of sterile statistics. Hiroshima was, in this retelling, the site of just another nameless, faceless, interchangeable, and unavoidable 100,000 casualties. Such is the nature of modern war, he asserted.

"[I]n Mr. Stimson's view," noted the *New York Times* editorial, "it is necessary to indict, not the weapons, but rather war itself."

But the post-bomb and post-"Hiroshima" anxiety and outrage had everything to do with the new weapons themselves: what they did to human beings, what they did to entire cities, their lingering lethalness, and what they portended for humanity's future. Stimson's article did not mention the words "radioactivity" or "radiation" once, but after Hersey's "Hiroshima" there would never again, despite best efforts to the contrary, be a successful return to the portrayal of the atomic bomb as a conventional weapon. No longer could reports of atomic bomb radiation poisoning be dismissed as "Japanese propaganda" or "Tokyo tales." That effort on spin had failed too; this aspect of the story of the bomb had now permanently gotten away from the government.

If Stimson, Conant, President Truman, and General Groves regarded "the face of war" as an anonymous abstraction, for millions of Hersey's readers, the face of atomic war had now been connected with actual human faces: those belonging to a struggling widow and her three children, a young clerk, two doctors, a priest, and a pastor. Photographs of those six faces of war covered the back of the book now selling out around the globe. *Hiroshima* had become a shadow that would forever cast doubt on the official justifications for the bombings.

Hersey's book continued to sell out around the world. Just two weeks after the Stimson article was released in *Harper's*, Hersey and the *New Yorker* editors were informed that Penguin Books in the United Kingdom had sold out of its quarter million first printing within weeks and was preparing to release a new printing of 1 million copies. A year after Knopf first released the book in the United States, the publisher issued a press release summarizing the book's worldwide reach: "Besides [the Knopf] American edition, and the British Empire's Penguin Books edition, 'Hiroshima' has appeared in eleven languages: Swedish, Danish, Norwegian, Finnish, Dutch, French, Czech, German, Italian, Hungarian, and Portuguese, and is scheduled to appear before long in Polish, in Spanish, and Hebrew. It is also likely that the book will appear in the Bengali and Marathi languages of India." There was even a Braille edition.

FALLOUT

In the fall of 1946, James Conant had fretted that the post-"Hiroshima" criticism would influence the country's entire next generation of historians and leaders, and his fears were immediately realized. Hersey's book was not only deemed an instant classic by booksellers and critics but also quickly made its way into college curriculums—including some of the Ivy League universities in the Harvard president's community. (Upon learning that "Hiroshima" was being made into a textbook edition for use in high schools, Ross wrote to Hersey and Shawn, "Christ knows what we have wrought. It will make a better text book than they had in high schools when I was there, at that.") *Hiroshima* had become the document of record about the true human cost of nuclear war, and was destined to remain so for decades.

THE DOUBLE AGENT

There were two notable absences on Knopf's brag list of countries where *Hiroshima* had been published. When "Hiroshima" was first released, the *New York Herald Tribune* had predicted that—despite the story's global reach—there would be millions of people who would never be able to read Hersey's account, thanks to repressive regimes around the world and the censorship they imposed on their populations.

The editorial was clearly implicating Russia. Every week, the Cold War was escalating further. Two months after "Hiroshima" was released by the *New Yorker*, the Soviet foreign minister, Vyacheslav M. Molotov, gave the United States a dramatic dressing-down in a speech to the General Assembly of the United Nations. He chided America's "imperialist" and "expansionist plans" and informed the Americans that the Soviet people had not just "shed their invaluable blood in streams to pave the way for new claimants to world domination." Furthermore, Molotov had accused the United States of selfishness in its "monopolistic possession of the atomic bomb." That atomic monopoly, Molotov warned, could not last long.

Soviet leader Joseph Stalin had indeed accelerated the Soviet Union's own atom bomb development program immediately after the United States had tested its first atom bomb in July 1945. With the Japan bombings, President Truman wanted to shock the Soviets, "to show who was boss," thought Molotov. To him and other Soviet leaders, the Hiroshima and Nagasaki bombs "were not aimed at Japan but rather at the Soviet Union," Molotov contended. "They said, bear in mind you don't have an atomic bomb and we do, and this is what the consequences will be like if you make the wrong move."

The issue of trying to get "Hiroshima" released in the Soviet Union had come up in the *New Yorker* offices even though Hersey, Shawn, and Ross knew that their article would likely be interpreted as a propagandistic threat by the Soviets. Some American editorials immediately drew attention to this fact: the *New York Herald Tribune* predicted that the Soviets would see Hersey's article as an "attempt to intimidate them."

The Soviets had actually gotten into Hiroshima to inspect the aftermath of the atomic bombing sooner than the Americans: the Soviet Union had not declared war on Japan until August 8, 1945—two days after the bombing of Hiroshima—and that country had maintained an embassy in Tokyo during the war. On August 23, Mikhail Ivanov, consul at the Soviet embassy in Tokyo, went to Hiroshima to inspect the aftermath. A quiet report on the horrific devastation and aftereffects of the bomb was quickly prepared and submitted to Stalin and other Soviet leaders. It was official: the Soviets were now at a terrible disadvantage vis-à-vis the United States. The Soviet government quickly suppressed press accounts about this devastating weapon, because, as one British Moscow-based correspondent noted, it had become clear in that city that "the bomb constituted a threat to the Soviet Union." If a story like "Hiroshima" were to be released in Russia, it would undermine that government's efforts to downplay the power that America now wielded over that country—and the world.

Still, the possibility of having the story published in Russia proved a tantalizing challenge to the *New Yorker* team, still giddy over its international triumph. They had just achieved a seemingly impossible feat together; perhaps they'd do it again. After some discussion, they decided that the team would contact Andrei Gromyko, the Soviet Union's ambassador to the United Nations, about the possibility of getting "Hiroshima" translated into Russian and distributed in that country. The outreach to Gromyko would have to be worded with delicacy, in hopes of setting the Russians' minds "relatively at ease as to possible propagandistic tendencies in the article," Hersey thought. (He likely suspected that nothing about "Hiroshima" would set the Russians' minds at ease, however. He himself had written, during his tenure as a Moscow correspondent for *Time* magazine, that "not a word is written in Russia which is not a weapon.")

Hersey, Ross, Shawn, and the *New Yorker*'s publisher, Raoul Fleischmann, spent a week drafting an entreaty to Gromyko, whose nicknames around New York included "Mr. Nyet," "Grim Grom," and "Old Stoneface." Their proposal was predictably met with stony silence. Furthermore, just months after their outreach to the Soviet ambassador, the *New Yorker* team would learn the reality of how Hersey and his story were regarded in Moscow—and just how naïve their bid for Russian translation and distribution had been in the first place.

Not long after Hersey's trip, a Soviet journalist named Oskar Kurganov arrived in Japan, dispatched by *Pravda*, the USSR's most prominent daily newspaper and the official mouthpiece of the Communist Party. He had toured the country, including a trip to Nagasaki. (He later likened his SCAP minders on this trip to the "American version of the Gestapo.") Upon his return to Russia, Kurganov wrote and released a book—*Amerikantsy v Iaponii*, or *Americans in Japan*—describing his Japan tour. Scenes such as those reported in "Hiroshima" had, according to Kurganov, been wildly exaggerated. Nothing like "atomic fever" existed. He reported that he had quizzed Nagasaki medics about the supposed affliction; they allegedly told him that they'd never seen a single case of it in that city. Kurganov also claimed to have interviewed a man who had survived the blast just by sheltering in a shallow ditch; this interviewee had even had his head exposed to the blast and sustained no injuries, just a "little fright." Kurganov stated that he personally was convinced that there was no such thing as radiation poisoning and that "no such 'atomic tragedies,' as were described to us by Americans, took place in Nagasaki."

The Kurganov book was a telling "Russian reply to the picture of destruction painted by John Hersey in 'Hiroshima,' " as one Moscow-based Western reporter noted. It was, in fact, essentially the Russian anti-"Hiroshima." Kurganov's message to Soviet readers: they need not fear the atomic bomb any more than America's other bombs; the United States had been lying about its new weapons and had no significant military advantage after all.

At the same time, *Pravda* released an article directly attacking Hersey. His "Hiroshima" was nothing more than an American scare tactic, a fiction that "relishes the torments of six people after the explosion of the atomic bomb," the story stated, which added that Hersey's motive was to "spread panic." (The article also stated that the book had sold over 7 million copies, implying that Hersey—a capitalist writer—must be getting quite rich off the sufferings he depicted.) In another Soviet publication, Hersey was called an American spy who embodied his country's "military spirit" and had helped to inflict upon the world a "propaganda of aggression, strongly reminiscent of similar manifestations in Nazi Germany."

Hersey and "Hiroshima" were now Cold War pawns. One country's maverick whistle-blower was another's virulent propagandist. If there was a silver lining for the U.S. government during the furor following the initial release of "Hiroshima" in the *New Yorker*, it may have been the knowledge that Hersey's story would cause great discomfort to their Soviet adversaries.

In the Soviet Union, any journalist who somehow managed to embarrass the government likely could expect to face dire consequences. It does not appear that the U.S. government or military interrogated or attempted to discredit Hersey or his reporting, nor tried to impugn the reported testimonies of the six "Hiroshima" protagonists. Theirs seems to have been a tactic of trying to downplay the story or attribute remorse over the human cost of the Hiroshima and Nagasaki bombs to oversentimentality. That said, even if there had been any real effort to discredit "Hiroshima," Hersey and the *New Yorker* editors had an ace up their sleeve: they could have simply revealed that General Groves and his aides had read and provided input on the story, then cleared it for publication.

However, several years later, in 1950—with the Cold War in full swing and Senator Joseph McCarthy fanning fears of a domestic "Red Scare"—FBI director J. Edgar Hoover assigned field agents to research, monitor, and interview Hersey, on whom the Bureau had already been keeping a file. The official reason given for this investigation: in 1941, Hersey's brother had been involved with an organization cited by the House Un-American Activities Committee as a Communist front. During their investigation, sources told FBI investigators that, during his Moscow posting for *Time*, Hersey had been "obviously and quite openly favorable to the Union of Soviet Socialist Republics" and that, once back in the U.S., he became involved with or made financial contributions to organizations with Communist links or sympathies. (This included a $10 contribution to the American Civil Liberties Union.) Of special interest to the FBI was a May 18, 1945, lecture Hersey gave at Yale in which he called for a "strong and lasting" friendship between the United States and the Soviet Union.

Hersey and his family had since moved to suburban Connecticut; FBI officials came to interview him at his home. He was questioned about his 1946 Japan trip to probe his relationships with other reporters there who may have had Communist sympathies.

Hersey was apparently not flagged for further interrogation; nor were any charges brought against him. It is probably unsurprising that Hersey's loyalties and background would be questioned in that charged political landscape, especially given his willingness to report blockbuster stories damaging to the U.S. government's reputation. Yet it is ironic that the FBI investigated Hersey as a possible pro-Russia Communist sympathizer, given that, in the Soviet Union, he had been branded a militaristic American spy determined to sow fear in Russia and around the world.

A FINE SPIRIT OF CHRISTIAN BROTHERHOOD

The second notable country in which "Hiroshima" was banned: Japan. Even if General MacArthur intended to use "Hiroshima" as training material among his own Pacific theater forces, Hersey was informed by a Tokyo-based *Life* photojournalist that SCAP was blocking both the article and the book version from being reprinted or distributed in Japan, or translated into Japanese.

Even though millions of readers around the world now knew their names, Hersey's six protagonists had to wait months before they could read the story. Hersey tried to mastermind ways to get copies of the *New Yorker*'s "Hiroshima" issue to Reverend Tanimoto, Miss Sasaki, Dr. Fujii, Mrs. Nakamura, Father Kleinsorge, and Dr. Sasaki. The *New Yorker* team sought guidance from the New York chapter of Jesuit Missions, an international Jesuit organization, and were told by a reverend there to send copies via a chaplain who was presently in Tokyo. The top copy in the stack, the organization rep advised, should be marked just for the chaplain, as a SCAP censor would likely confiscate copies earmarked for Japanese nationals.

Eventually copies of the issue—whether via this chaplain or someone else—did reach Hiroshima. Some of Hersey's protagonists were not aware that they had been included in the article—or that the article had even been written in the first place—until they received contraband copies of the magazine. "I didn't know that he had put me in his story until someone came running over with a copy of the *New Yorker* to tell me about it," Dr. Fujii recalled later.

Upon reading it, Dr. Fujii found that "everything in [it] was just as he said it was" and that Hersey had remembered every word of their three-hour conversation. (Mrs. Nakamura also marveled at Hersey's ability to "remember all the tiny details.") Dr. Fujii wrote Hersey a postcard in which he extended his "best thanks for your kind present of 'New Yorker'" and added, "Suppose how I am pleased to read the article 'Hiroshima' about A-bomb especially about me myself. I believe your kind expression would have caused a big sensation in the world."

A letter also eventually arrived from Reverend Tanimoto, who told Hersey that he had been "greatly surprised and excited" by the story— and stunned that the *New Yorker* had been allowed to publish it.

"It is a marked and rather cynical contrast to the former leaders of Japan as compared with the American authorities, who gave you permission to publish such a report on the effects of the secret weapon as viewed from the standpoint of the defeated nation," he wrote. "We now witness the fine demonstration of the American Democracy and the strong sense of humanity and the fine spirit of Christian brotherhood in the sensational reaction of your people."

He reported that he and other "Hiroshima" protagonists had begun meeting for monthly reunions—they called their gatherings the "Hersey Group," Reverend Tanimoto reported—and added that he had heard that SCAP had still not approved the article for translation into Japanese. If anything changed on that front, he volunteered to undertake the job.

Last but not least, Reverend Tanimoto informed Hersey that the U.S. commander of the Eighth Army had recently visited Hiroshima and "interviewed" him, Father Kleinsorge, Dr. Fujii, Dr. Sasaki, and Miss Sasaki. (He hadn't seen Mrs. Nakamura since the visit, so he didn't know if she, too, had been sought out and questioned by the Army commander.)

It would take more than two years and the intervention of the Authors' League of America to secure permission from General MacArthur to have the book distributed in Japan. Upon greenlighting the translation at last, the general stated that reports that he had blocked the "Hersey book" were "without the slightest basis in fact."

"They stem from a maliciously false propaganda campaign," he added, "aimed at producing the completely fallacious impression that an arbitrary and vicious form of censorship exists here."

Upon its April 25, 1949 release, the Japanese edition of *Hiroshima*—co-translated by Reverend Tanimoto—became an immediate bestseller. If SCAP had worried that the book might stir up feelings of bitterness or retribution—or otherwise "disturb the public tranquility," as dictated in its occupation press code—Japanese reviewers seemed to regard *Hiroshima* with a mix of sadness and cautious optimism. It "expresses a humanism which transcends the positions of victor and vanquished," wrote one Japanese reviewer in the *Tokyo Shimbun*. "It should be read seriously, with poignant hope for peace."

I DON'T CARE WHAT THEY SAY ABOUT ME

That following spring, in 1947, Hersey's editors at the *New Yorker* were trying to devise their next Hersey collaboration. Following up their major reporting coup would be no small task. Ross had been flirting with the idea of sending Hersey back to Japan for a follow-up story. "Hold a reunion with your characters, and report it," he suggested to Hersey.

Hersey shot the idea down. He had essentially gone underground, journalistically speaking, for months after "Hiroshima" was released and showed no inclination to go back to Japan right away. The odds of getting Hersey to return were "probably about forty-two to one," Ross lamented to a contact at the North American Newspaper Alliance.

Now back in the United States with his wife and family after years of war reporting abroad, Hersey began to shift his attention from reporting to fiction, acting on his theory that fiction could be stronger and more affecting than nonfiction. (His theory would prove ironic, given that "Hiroshima" remains his best-known and most influential work—despite the fact that Hersey authored more than a dozen novels throughout his lifetime.) He had begun to research a novel, set in the Warsaw Ghetto, whose ruins he had toured when still a Moscow-based war correspondent for *Time* magazine. The book would eventually be titled *The Wall*. As when bearing witness to the carnage in Hiroshima, Hersey had been devastated by the ghetto ruins and concentration camps he had seen. After that trip, it had taken him a long time to process his "outrage at human capabilities." Yet "the experience gave rise to certain optimism, too, for in each case there were survivors, and one had to conclude that mankind is indestructible," he stated later.

He emerged from his project research for *The Wall* to do a profile for the *New Yorker* on Bernard Baruch, who had then just resigned as the U.S. representative on the United Nations Atomic Energy Commission. The profile ran in January 1948, nearly a year and a half after "Hiroshima" had come out. After that, the "Hiroshima" editorial trinity—Hersey, Ross, and Shawn—would collaborate on only one more major journalism project: a five-part profile of President Truman that ran in early 1951. Harold Ross would die in surgery later that year after a struggle with cancer.

It was something of a miracle that the team was able to wangle access to the president after the damage caused by "Hiroshima," but Ross had been busily, craftily cultivating the president's press secretary, Charles Ross, in the years since. In making the pitch to President Truman's team, the *New Yorker* editors used an argument that was an ironic riff on their approach to "Hiroshima." Their objective "would be to give . . . a picture of Truman as a Human Being," as Shawn wrote to Ross when instructing him on how to approach the White House with the proposition. The tactic worked: Hersey was given substantial access to the president.

When Hersey shadowed and interviewed President Truman in late 1950, the Soviets had just successfully detonated their first atomic bomb the previous year, forever ending the America's reign as nuclear monopoly holder (and proving wrong General Groves's speculation that it might take the Soviets as many as twenty years to join the nuclear club). President Truman had immediately vied to regain the advantage, accelerating U.S. efforts to create thermonuclear weapons. In 1952, the United States would successfully detonate its first hydrogen bomb—the "Hell Bomb," as Atomic Bill Laurence called it—with a payload equivalent to more than 10 megatons of TNT, making it approximately 666 times as powerful as the Hiroshima bomb.

If Hersey and President Truman spoke about these matters, or Hiroshima and Nagasaki, those conversations did not make it into the published profile in the *New Yorker*. President Truman had rarely spoken in public on the subject of the Japan bombings. The *Atlantic* had since published the president's letter to physicist and Interim Committee member Karl Compton—in which the president had stated that the "Japanese were given fair warning"—and this 98-word missive was about as elaborate a statement as the American public had gotten about the president's personal viewpoint on the bombings. It is unclear if making the topic of the bomb off-limits was a White House precondition for agreeing to the *New Yorker* profile, but Hersey did attempt to venture into those waters, asking President Truman if there is "anything you'd like to add to the list of ten books to prepare a man specifically for life in the Atomic Age." The president declined to offer such a list other than telling Hersey to refer to the classics.

"There's nothing new in human nature," Truman told him. "Only our names for things change."

Hersey was allowed to attend and document a meeting in which President Truman and his aides worked on a radio speech that Truman reportedly "regarded as the most important one of his career up to that time." Chinese Communists had been making headway in Korea, and the president was about to declare a state of national emergency. In the meeting, President Truman's team of advisors discussed intelligence on the matter gathered in Tokyo; Japan had indeed become the reliable Pacific theater foothold the United States had hoped for when it had occupied that country six years earlier. In Hersey's depiction of the meeting, the president and Secretary of State Dean Acheson debated language for portraying the Soviets as responsible for bringing the world to the verge of nuclear war.

"I don't think we should leave any doubt in anyone's mind that we will never start a world war—that if it starts, it will have to be the other fellow who starts it," Acheson reportedly said. "Perhaps we should say something like [the Communists] have shown that they are willing to bring the world to the brink of war.'"

"That's just what the Russians are doing," President Truman responded. "Pushing everybody."

At that time, in a White House press conference about Korea—also covered by Hersey for the *New Yorker*—President Truman told reporters that use of the atomic bomb against the Chinese Communists was not out of the question. Moreover, Hersey wrote, the president informed the White House press corps that "the government had exerted every effort to prevent a Third World War [and] was still trying to prevent that war from happening."

Here was a portrayal of an American president who had inherited the last world war in its final stages, and was now navigating the early stages of the potential next one. If President Truman evinced to Hersey in their private interviews any wariness over the current course, or second-guessed his role in the atomic bombings of Hiroshima and Nagasaki, Hersey did not report it in the *New Yorker* profile. President Truman did admit to Hersey, however, that he was bothered by some assessments of himself in the press—which occasionally bordered on the treasonous, in his opinion. "To my mind, there's nothing as un-American as a lying smear on a man's character," he said. Here the president wavered between defiance and vulnerability.

"I don't care what they say about me," he told Hersey. "I'm human. I can make mistakes. Any man can make mistakes, even if he's trying with all his heart and mind to do the best thing for his country."

Throughout the remainder of his presidency, Truman would always keep atomic weapons in "active consideration" for use in military situations that might arise on his watch. They were no different, he stated, from conventional weapons, just bigger, more efficient, and more effective—a legitimate part of the U.S. arsenal. His successor, President Dwight D. Eisenhower, felt much the same way, even reasoning that the atomic bomb was a valuable cost-saving device. It might be cheaper to use atomic weapons against North Korea than using conventional weapons, he stated in a May 13, 1953, National Security Council meeting. Not to mention that using nuclear weapons would save the logistical costs and bother of getting conventional ammunition from the United States to the front lines in that country.

HERE TO STAY

With the nuclear arms race escalating in the years and decades that followed the release of "Hiroshima," Hersey had an evolving and complicated view about the legacy and influence of his story. Decades later, in the 1980s, against the backdrop of renewed, fever-pitch nuclear brinksmanship between the Soviets and Americans, he told one scholar that he felt that "the bombings of Hiroshima and Nagasaki had been a significant warning to the world, and had contributed to preventing another nuclear war." He also thought that bomb survivor testimonies, such as the ones he had relayed in the *New Yorker,* had had a significant impact.

"I think that what has kept the world safe from the bomb since 1945 has not been deterrence, in the sense of fear of specific weapons, so much as it's been memory," he said in 1986, in a rare interview. "The memory of what happened at Hiroshima." As long as people remembered vividly what had happened there, and at Nagasaki, they could still imagine "what it would have been like to have a much bigger bomb dropped on a center of population"—and what it would be like if their own cities and children were the targets of nuclear attack—and advocate against future use of atomic weapons.

Still, he was deeply concerned: while "Hiroshima" had influenced subsequent generations of political leaders, activists, and academics working to curb or end the nuclear arms race, the memory of the bombs' aftermath appeared to be growing "very spotty in the centers of power" in Washington, D.C. Hersey cited figures such as then Secretary of Defense Caspar Weinberger and Assistant Secretary of Defense for Global Strategic Affairs Richard Perle, who "must never have grasped the meaning of the Hiroshima bomb, the way they go on about a future with bigger and better nuclear weapons." He emphasized that in the Soviet Union there was probably no widespread institutional memory at all: "The control of information there is such that I wonder how many really know what happened at Hiroshima."

The fading of memory—or lack of memory—was, in Hersey's opinion, a true threat to deterrence, making the testimonies of Reverend Tanimoto, Mrs. Nakamura, Miss Sasaki, Dr. Fujii, Father Kleinsorge, and Dr. Sasaki more crucial than ever.

He personally saw the six "Hiroshima" protagonists—as well as the many other war survivors he documented over the years—not just as cautionary tales but emblems of hope. Throughout his life and career, Hersey remained fascinated by the human will and ability to survive, and was almost shockingly optimistic for a man who had witnessed firsthand so much of the worst in human nature. Man has "astonishing resources for holding on to his life," Hersey wrote.

"In spite of the appalling tools [man] invents to destroy himself," he concluded, "it seems to me that he loves this seamy world more than he desires, as he dreads and flirts with, an end to it . . . I believe that man is here to stay."

Epilogue

Even though Harold Ross had hoped that John Hersey might return to Japan soon after "Hiroshima" came out, it would take nearly four decades for Hersey to go back to that country and report again on the fates of his six protagonists. In 1985, the *New Yorker* published a new "Reporter at Large" story by Hersey: "Hiroshima: The Aftermath." Following Ross's death in 1951, William Shawn had become the longtime editor of the magazine (he was, the *New York Times* would write, the *New Yorker*'s longstanding "gentle despot"), and he oversaw Hersey's sequel story.

"It was quite clear that the shadow [of Hiroshima in the lives of his six subjects] was longer than the one year about which I had written in the original piece," Hersey recalled. "The shadow was much, much longer." He decided at last to go back and investigate what had happened to them and their families.

In the intervening years Hersey had continued to keep relatively quiet about his 1946 trip to Hiroshima. He gave few interviews on the subject—or any other subject, for that matter. Unlike many of the hotshot journalists who followed in his footsteps, such as Norman Mailer and Tom Wolfe, who were often prominent in their own stories and reveled in their celebrity, Hersey had continued to shun publicity, writing over two dozen novels and nonfiction books from his homes in suburban Connecticut, Martha's Vineyard, Key West, and New Haven, where he was master of Yale's Pierson College for half a decade. "He was a member of the generation that developed the cult of the author—people like Norman Mailer were doing *The Dick Cavett Show*—but he didn't want any part of that," recalled his son, Baird Hersey. "He never went on tour . . . He didn't go on TV or radio, didn't give lectures." On the subject of the celebrity journalists who were succeeding him, Hersey called himself "one worried grandpa" and added that "the figure of the journalist [was becoming] more important than the events being written about."

Upon arriving in Hiroshima nearly forty years after Little Boy had decimated it, Hersey found that a "gaudy phoenix had risen from the ruinous desert of 1945." The rebuilt city now contained more than a million inhabitants. Trees lined the new broad avenues. Hiroshima was now a "city of strivers and sybarites," he observed, adding that there were hundreds of bookstores and thousands of bars.

Nearly every August 6, after the occupation ended, press outlets from around the world had sought Hersey's six "Hiroshima" survivors out for anniversary interviews. In the late 1940s, Reverend Tanimoto had become an internationally known antinuclear advocate. In the years following the bombing, the pastor began traveling to the United States to raise funds to rebuild his church and speak about his experiences, and also help spearhead an effort to secure reconstructive surgeries for young Japanese women disfigured in the atomic bombing. By one account, between 1948 and 1950, he gave 582 lectures in the United States before returning to Japan. During a second tour, on February 5, 1951, he was invited to give the opening prayer for the afternoon session of the U.S. Senate. In the prayer, Reverend Tanimoto described America as "the greatest civilization in human history" and added, "[w]e thank Thee, God, that Japan has been permitted to be one of the fortunate recipients of American generosity." The appearance, Hersey reported, was "the high point of the trip (and possibly of [Reverend Tanimoto's] life)."

A lower point, however, occurred in 1955. While on another trip to America, Reverend Tanimoto was invited to do a May 11 television interview with NBC in Los Angeles. When he arrived on the set, it turned out that he was making an appearance on *This Is Your Life*, broadcast to around 40 million Americans. As Reverend Tanimoto sat there on the set, cameras whirring, the show's host, Ralph Edwards, said to the pastor, "You thought, of course, you were going to be interviewed as part of the work you are now doing, didn't you? We may have a little surprise." He then informed Reverend Tanimoto that they would "retell the story of your life . . . on this stage. We hope you have some pleasant moments." He then asked the shocked minister to relive his experience of August 6, 1945; as he spoke, sound effects blared in the background, including air-raid sirens, swells of faintly Asian-sounding music, a ticking clock, and a terrific explosion. Reverend Tanimoto's shotgun testimony was interrupted for an interminable on-set commercial demonstration for one of the show's sponsors, Hazel Bishop nail polish, during which a hand model zealously scrubbed her polished nails with steel wool to demonstrate the product's tenacity. Reverend Tanimoto was then asked to continue his remembrance about surviving a nuclear apocalypse.

The show producers had not only secretly flown Reverend Tanimoto's family over from Japan—including his wife, Chisa, and daughter, Koko, both blast survivors—they also trotted out Captain Robert Lewis, copilot of the *Enola Gay* on the mission that had dropped the bomb on Hiroshima. Captain Lewis had since been working as a personnel manager of a New York–based candy-making company. At one point, it appeared that Lewis was starting to cry as he recounted the dropping of the bomb from his B-29; Koko Tanimoto, then ten years old, also saw tears in the man's eyes and reached out for his hand despite her feelings of hatred upon first seeing him. (Hersey reported in "Aftermath" that Lewis had actually not been crying; rather, "he had gone out bar-crawling" before the show and was drunk.) There was some solace to be found in Reverend Tanimoto's *This is Your Life* ordeal; the show reportedly brought in around $50,000 in donations from viewers.

Reverend Tanimoto retired in 1982 and died in 1986 in a Hiroshima hospital of pneumonia complicated by kidney failure. He was seventy-seven years old.

Father Wilhelm Kleinsorge was also often approached for interviews in the years after "Hiroshima" was released. He made media appearances on German radio and television shows, although he eventually became a Japanese citizen, and took the name Father Makoto Takakura. "When Hiroshima was destroyed, I became determined to become Japanese," he told one interviewer. "I want to remain forever in Hiroshima, as an instrument of God's will."

He suffered from considerable health ailments throughout his life— including infection, an "A-bomb cataract," and chronic flulike symptoms—and died in 1977. On his hospital chart in 1976, a hospital worker had written "a living corpse." Bedridden at the time of his death, he told his caregiver—a Japanese woman named Satsue Yoshiki—that he read the Bible and timetables, because they were "the only two sorts of texts . . . that never told lies," Hersey reported.

Dr. Masakazu Fujii reveled in the fame bestowed upon him by "Hiroshima," although he also conceded that at times the attention could be overwhelming. "After being written about by Hersey, every year around [the bombing's anniversary] there's a lot to do and it's a bit inconvenient," he told an interviewer in 1952. It had taken him years to recover from the ordeal—"I struggled emotionally, physically, materially," he said—but he was, at least, able to reestablish his practice quickly. Just a few years after the bombing, an American doctor passed by his new hospital—which Dr. Fujii had rebuilt on the site of the collapsed old one—and noticed a sign in English that stated:

HERE IS FUJII, OF HERSEY'S HIROSHIMA FAME

By 1951, the sign had been upgraded to read:

DR. FUJII HERE
DR. FUJII, ONE OF THE SIX IN THE WORLDWIDE
KNOWN AS HIROSHIMA BY JOHN HERSEY
CAME BACK TO THE ORIGINAL
PLACE HERE AFTER A LAPSE OF
THREE YEARS SINCE THE A BOMB

The doctor remained proud of his affiliation with Hersey and about having been included in "Hiroshima." He kept Hersey's business card in his wallet and produced it proudly for visitors. "It's become one of my prized possessions," he later declared. During the occupation years, Dr. Fujii's practice had thrived. He lived comfortably and joined a country club.

Upon Dr. Fujii's death in 1973, the American-run Atomic Bomb Casualty Commission (ABCC)—first established in Japan during the occupation to study the aftereffects of the bombings—performed an autopsy on the doctor, revealing "a cancer the size of a Ping-Pong ball in his liver."

In the years after the bombing, Dr. Terufumi Sasaki remained at Hiroshima's Red Cross Hospital, with much of his work involving the removal of keloid scars covering many of his blast-survivor patients. He then opened up a private clinic and, like Dr. Fujii, became prosperous, but struggled at times with being in the limelight as a "Hiroshima" protagonist. "There was a mountain of letters from America," Dr. Sasaki recalled later. At first he attempted to reply to the correspondents but eventually stopped. "I don't want to think about that time anymore," he told a Japanese interviewer.

"For four decades, he almost never spoke to anyone about the hours and days after the bombing," Hersey reported in his follow-up *New Yorker* story. "[H]is one bitter regret: that in the shambles of the Red Cross Hospital in those first days after the bombing, that it had not been possible . . . to keep track of the identities of those whose corpses were dragged out to the mass cremations, with the result that nameless souls might still, all these years later, be hovering there, unattended and dissatisfied."

Miss Toshiko Sasaki in 1947 began working in an orphanage in Hiroshima. Like Hersey, she was stunned by the rapid rebuilding of the city upon the ashes. "I wouldn't say the city is being reconstructed so much as it's a completely new city," she said in the 1950s. She had undergone three more major surgeries on her leg within fourteen months; after this, she could walk almost normally. In 1954, with the guidance of Father Kleinsorge, Miss Sasaki entered a convent and in 1957 took vows and became Sister Dominique Sasaki.

She continued to suffer from fevers, blood spots, night sweats, and liver dysfunction—"a pattern of ailments which—as with so many hibakusha—might or might not have been attributable to the bomb," Hersey reported. Like Dr. Sasaki, she preferred not to discuss the events of August 6, 1945.

"It is as if I had been given a spare life when I survived the A-bomb," she said. "I shall keep moving forward."

Mrs. Hatsuyo Nakamura continued to be wracked with illnesses in the post-bomb years but was too poor to consult doctors. (The Japanese government would not offer meaningful medical assistance to hibakusha until 1957.) She remained in Hiroshima and survived by doing odd jobs, including delivering bread for a bakery (earning about 50 cents a day), peddling sardines from a street cart, and collecting money for deliverers of the local newspaper. She eventually found long-term employment at a company that produced mothballs. All three of her children graduated from school and married. There is evidence that at least one of them suffered from PTSD after the bombing: Mrs. Nakamura later told a reporter that her daughter Myeko was "so afraid of war after being buried up to her chest when our house collapsed, I thought about evacuating [from Hiroshima] to the mountains."

William Shawn was ousted as editor of the *New Yorker* in 1987 by S. I. Newhouse Jr., whose Condé Nast Publications acquired the magazine in 1985. Shawn died of a heart attack in 1992.

John Richard Hersey died of cancer a few months later, on March 24, 1993. He was seventy-eight years old.

Today, Hiroshima Prefecture has nearly 3 million inhabitants. It maintains a world-class museum documenting the atomic bombing and its aftermath, as well as a park and many monuments. The Atomic Bomb Dome—a building whose structure partially survived despite being located near the bomb's hypocenter—is a UNESCO World Heritage Site.

Located today at the site of the bomb's exact hypocenter: a low-rise medical building and a 7-Eleven convenience store.

* * *

According to a survey conducted soon after the release of "Hiroshima," the majority of those surveyed viewed Hersey's story as a contribution to the public good. "Hiroshima" had indeed been written out of concern for the future of humanity in its entirety, not just the welfare of one nation or race or political party.

The article also served as an unnerving reminder to readers that their elected government leaders operated on many clandestine levels, and not always in their best interest. Hersey and his *New Yorker* editors created

"Hiroshima" in the belief that journalists must hold accountable those in power. They saw a free press as essential to the survival of democracy—a form of government which had just narrowly escaped extinction.

For Hersey and many of the Allied correspondents who had covered the second world war, the global conflict had been a battle to preserve such ideals. For *Chicago Daily News* reporter George Weller, first foreign correspondent into post-bomb Nagasaki, the American people had been "fighting to be informed," he stated.

"They did not want to be fooled," he said. "They wanted to hear the truth. They could take it."

Before reporting and writing "Hiroshima," Hersey had satirized despotic American general George S. Patton in *A Bell for Adano* because the military leader's arbitrary cruelty epitomized for Hersey "the very things we were fighting against." Journalists like Hersey and Weller would not be muzzled, they declared, nor were they willing to allow those in power to denigrate the freedoms that had just been so painfully defended.

In 1937, before joining the staff at *Time* magazine, Hersey had worked as an assistant to writer Sinclair Lewis, who had, in his 1935 novel *It Can't Happen Here*, warned Americans that what they were seeing happen in Europe—the rise of toxic populism and of vicious government propaganda machines, the assault on truth and facts, the ascent of despotic leaders—could indeed happen in the United States, even though Americans tended to see themselves as almost compositionally invulnerable to such events. Hersey and others of his generation can be forgiven for hoping that the outcome of World War II—with democracy triumphant—had proved Lewis wrong.

Yet the greatest tragedy of the twenty-first century may be that we have learned so little from the greatest tragedies of the twentieth century. Apparently catastrophe lessons need to be experienced firsthand by each generation. So, here are some refreshers: Nuclear conflict may mean the end of life on this planet. Mass dehumanization and racism can lead to genocide. The death of an independent press can lead to tyranny and render a population helpless to protect itself against a government that disdains law and conscience.

If American civilians in 1945 were exhausted by years of demoralizing wartime news, many Americans today feel comparably overwhelmed and debilitated by events of recent years, and the sheer volume of news, information, and disinformation inundating them every day. But succumbing to numbness and indifference will have disastrous consequences. No matter how exhausting and daunting the landscape, it remains imperative for Americans not to try to "escape into easy comforts" again, as Albert Einstein put it. The Trump-era assault on the press in the United States would have appalled Hersey and his colleagues, who likely would have seen the current war on facts as one of the most alarming challenges of our time, if not an existential threat in its own right.

At this moment of reckoning, Americans must honor and fiercely defend the fourth estate and its pursuit of truth in this country. The opportunity to learn from history's tragedies has not yet passed.

Acknowledgments

Researching and writing *Fallout* has been the greatest honor of my professional life, yet it was a herculean effort that required the help and support of many people. My editor, Eamon Dolan, and literary agents, Molly Friedrich and Lucy Carson, helped me conceive and shape this project from its earliest stages and were closely involved throughout its evolution. *Fallout* could have been 250 pages or it could have been 1,000; with each subsequent draft, Eamon shrewdly helped bring out the narrative from a mountain of material, and charmed his way through ordering often painful cuts to the manuscript. Molly and Lucy provided frequent, invaluable editorial input and encouragement.

Research for *Fallout* was conducted on three continents in four languages. I am particularly grateful to my primary Russian translator and researcher, Anastasiya Osipova, who became my general research associate, flanking me at several archives and acting as my emissary to others; pulling documents, articles, and many other materials for me around the clock; and acting as a crucial sounding board throughout the research, writing, and editing of this book. I also owe a great debt to my Tokyo-based researcher and translator, Ariel Acosta, who also acted as an assistant to me during my time in Japan. Deepest thanks to my German translator-researchers, Professor Sigi Leonhard and Nadja Leonhard-Hooper, and to my longtime research associate Alison Forbes for her document and article research as well as her help in tracking down elusive contacts. Michael G. Bracey provided essential research from the National Archives and Records Administration and patiently helped me and my team navigate the overwhelming SCAP, U.S. government, and U.S. military records there. Laura Casey helped fact-check the manuscript and provided important editorial input during the final stages of editing.

I was greatly honored to work with one of Hersey's last surviving "Hiroshima" protagonists: Koko Tanimoto Kondo, who personally guided me through Hiroshima, showed me the exact point of Little Boy's detonation, gave me several lengthy interviews, and offered significant insight into the hibakusha community in Japan. She has since become a treasured friend, and I have dedicated this book to her. I also deeply appreciate the help and support of descendants of other protagonists in *Fallout*, including Pater Franz-Anton Neyer, nephew of Father Wilhelm Kleinsorge; Kazue Tokita and Natsuko Tokita, daughter and granddaughter of Leslie Nakashima; Michael McCrary; George Burchett; Anthony Weller; Jennet Conant; and Leslie Sussan.

I am deeply grateful to Governor Hidehiko Yuzaki of Hiroshima Prefecture for his interview and support, and to Dr. Yoichi Funabashi, chairman of the Asia Pacific Initiative, for his expertise and introductions within Japan. I would also like to extend my appreciation to Professor Kazumi Mizumoto of the Hiroshima Peace Institute for his interviews and for patiently answering many questions. Thank you to Matt Fuller for his tireless efforts in making introductions in Japan on my behalf and other essential support and guidance. John V. Roos and William F. Hagerty IV, both former U.S. ambassadors to Japan, gave deeply appreciated and crucial interviews, and I am grateful to Brooke Spelman of the U.S. embassy in Tokyo and David Mandis, executive assistant to Ambassador Hagerty, for their assistance.

Several board members and associates of the Bulletin of the Atomic Scientists supported this project, and I am deeply grateful for their help and guidance. Former U.S. secretary of defense William J. Perry and former California governor Jerry Brown both gave important interviews. Thank you also to Dr. Kennette Benedict for her crucial input, and to Janice Sinclaire for her tireless and deeply appreciated support. My thanks also to Robin Perry and Deborah Gordon of the William J. Perry Project, and to Evan Westrup.

I am also grateful for the support I received from members of the *New Yorker*'s team, past and present. My thanks to *New Yorker* editor-in-chief, David Remnick, for his interview and encouragement, and to *New Yorker* writer Adam Gopnik, who was an essential consultant from the project's conception. I would also like to thank John McPhee, John Bennet, Bill Whitworth, Sara Lippincott, Jane Kramer, Anne Mortimer-Maddox, Martin Baron, Richard Sacks, and Pat Keogh for their recollections, references, and/or guidance. Thank you also to Natalie Raabe for fielding many questions throughout the research process, and to Fabio Bertoni for generously granting me permission to quote extensively from the *New Yorker*'s historical materials. Michael Gates Gill has my appreciation for his encouragement and information about his father's world at the *New Yorker* and Brendan Gill's friendship with Mr. Hersey, and to Susan Morrison for granting permission to use material from the Lillian Ross estate.

I was fortunate to have input from experts, biographers, and scholars in a wide array of fields throughout my research and writing. Thank you to Professor Martin Sherwin for his guidance on the Soviet Union and the bomb; to Dr. Robert Jay Lifton and Greg Mitchell for their interviews and help, and for their groundbreaking work on America's fraught relationship with Hiroshima; to Dr. Robert S. Norris for his essential assistance on information pertaining to General Leslie Groves; to Matt Korda, research associate of the Nuclear Information Project at the Federation of American Scientists (FAS), for his guidance on current nuclear weapon stockpiles worldwide and other historical technological facts about the atomic weapons detailed in this book; to Steven Aftergood, director of the FAS project on government secrecy; to Dr. William Burr, director of the National Security Archive's nuclear security documentation project; to Dr. Hans Kristensen, director of the Nuclear Information Project at FAS; and to Richard Rhodes for his guidance on Albert Einstein's role in the creation of the bomb.

Thank you also to *New Yorker* biographer Ben Yagoda and Harold Ross biographer Thomas Kunkel for their support and feedback, and to Gay Talese for his expertise on the historical *New York Times* and for making many introductions for me. Professor Michael Yavenditti has my gratitude for his support and for helping me track down his important 1970 dissertation, "American Reactions to the Use of Atomic Bombs on Japan, 1945–1947." Thank you also to the Foreign Correspondents' Club of Japan's Charles Pomeroy, a historian of foreign journalists in postwar Japan, for his help with many queries. I would also like to extend my appreciation to censorship historian Professor Michael Sweeney, First Amendment legal expert Jean-Paul Jassy, and film historian Professor Jeanine D. Basinger for her guidance on Japanese portrayal in World War II–era American commercial films, propaganda films, and military media materials. Thank you also to Professor Jian Wang, director of the USC Center on Public Diplomacy; to Willow Bay, dean of USC Annenberg School for Communication and Journalism, and to USC professors Joe Saltzman and Geoffrey Cowan; and to Professor Aleksandar Matovski, Professor Thomas Kohut, Professor Jim Shepard, and Professor Eiko Maruko Siniawer—political science, film, and historical experts at my alma mater, Williams College.

My deepest thanks to the many archivists who helped me and my team research this project, including Jessica Tubis and Anne Marie Menta of Yale University Library's Beinecke Rare Book & Manuscript Library; Virginia T. Seymour of the Harry Ransom Center at the University of Texas at Austin; Jeff Roth and Alain Delaqueriere of the New York Times Archives; James W. Zobel of the MacArthur Memorial Library & Archives; Jill Golden, director of the Life Photo Archive and Bill Hooper of the Time Inc. Archives; Eisha Neely of Cornell University's Division of Rare and Manuscript Collections; David A. Olson, Hong Deng Gao, and Thai Jones of Columbia University's Rare Book & Manuscript Library; Meredith Mann of the New York Public Library's Manuscripts, Archives, and Rare Books division; Sarah Patton and Diana L. Sykes of Stanford University's Hoover Institution Library & Archives; Tricia Gesner and Francesca Pitaro of the Associated Press Corporate Archives; Emma M. Sarconi and Gabriel Swift of Princeton University Library's Rare Books and Special Collections; Abigail Malangone and Matt Porter of the John F. Kennedy Presidential Library; Jeannie Rhodes, photography research director at *Vanity Fair*; Deirdre McCabe Nolan of the Condé Nast Library; Hiroko Moriwaki of the Foreign Correspondents' Club of Japan; Randy Sowell and David Clark of the Harry S. Truman Library & Museum; Bill Landis and Christine Weideman of Yale University Library's Manuscripts and Archives; the reference staff of the Harvard University Archives; and Yael Hecht of the Beverly Hills Public Library.

For *Fallout*, I was allowed to reference previously unpublished materials, and I would like to gratefully acknowledge those who gave me access to and permission to quote from them, including Koko Tanimoto Kondo for sharing her family's historical photos with me and allowing me to excerpt from her father's unpublished diaries and letters; Scott, Bonnie, and Peter D. Sanders for unearthing and sharing notes from an interview conducted with Mr. Hersey by their father, pioneering Hersey biographer David Sanders; and Leslie Sussan for sharing with me a portion of her unpublished manuscript detailing her father's time in post-bomb Hiroshima.

Acknowledgments

Several contemporary journalists and producers provided contacts, support, and guidance on war reporting culture, including Tom Bettag, John Donvan, and Jack Laurence. My thanks to David Muir for his guidance on reporting from nuclear catastrophe zones. The team from *PBS NewsHour*/Facebook Watch's *That Moment When*, including Sara Just, James Blue, Dana Wolfe, Steve Goldbloom, and Melissa Williams, have my deepest appreciation for devoting an episode to the evolution of *Fallout* and the continued significance of the events that the book documents. *Time* magazine's Brian Bennett provided crucial insight, support, and contacts; his wife, Anne Tsai Bennett, made important State Department inroads for me. I am grateful to them both. Thank you also to *PBS NewsHour* senior national correspondent Amna Nawaz; *PBS NewsHour* deputy senior producer for foreign affairs and defense Dan Sagalyn; conflict journalist and author Gayle Tzemach Lemmon and her husband, Justin Lemmon, for their support and crucial introductions on my behalf; Chip Cronkite for his early support; ABC News correspondents Karen Travers and Gloria Riviera; Kirit Radia of the ABC News foreign desk; Patrick Reevell, Moscow-based contributor to ABC News; Elizabeth Angell of *Town & Country*; David Friend of *Vanity Fair*; Anya Strzemein of the *New York Times*; and Nicole Rudick, former editor at the *Paris Review*.

Some of Hersey's former friends, students, and colleagues generously shared their time and recollections with me, including Rose Styron, Margaret Blackstone, Jane O'Reilly, Phyllis Rose, David Wolkowsky, Ross Clairborne, Lynn Mitsuko Higashi Kaufelt, and Phil Caputo. I am grateful to each of them. My thanks to Nathaniel Sobel for sharing his Yale dissertation on Hersey and his research materials with me.

I would also like to thank my team at Simon & Schuster, including Tzipora Baitch for her tireless support, guidance, and organization; copy editor David Chesanow for undertaking the massive, complex task of vetting this work, and for doing so with such patience; senior designer Lewelin Polanco; Stephen Bedford of Simon & Schuster's marketing team; legal advisor Felice Javit; production editor Kathy Higuchi; senior publicist Brianna Scharfenberg and publicity director Julia Prosser; and cover designer Rich Hasselberger. Many thanks also to my film agent, Howie Sanders of Anonymous Content, and his associate, Tara Timinsky, for their great support from the earliest stages of my research. My London-based literary agent, Caspian Dennis, and Henry Rosenbloom, *Fallout*'s editor at Scribe Publications, also have my greatest appreciation. Others who provided vital research support include Sasha Odynova, Moeko Fujii, and Annie Hamilton.

Thank you also to Lynn Novick; Sally Quinn; Masami Nishimoto of the *Chugoku Shimbun*; Dan Sloan of the Foreign Correspondents' Club of Japan; Sophie Pinkham; Liesl Schillinger; Glynnis MacNicol; Mark Rozzo; Andy Lewis; Van Scott Jr. and Julie Townsend of ABC News; Michelle Press of Getty Images; Harvey Jason of Mystery Pier Books, Inc.; Hugues Garcia; Emily Lenzner; Courtney Dorning; Dr. Jeffrey Neely; Katelyn Massarelli; Alexander Littlefield and Taryn Roeder of Houghton Mifflin Harcourt; Julia Demchenko; Melissa Goldstein; Heather Carr; Kent Wolf; Lorin and Sadie Stein; Herb Johnson and Lise Angelica Johnson; Ene Riisna and James Greenfield; Alex Ward; Sarah Rosenberg, Melinda Arons; Jin Pace; Gillian Laub; Lori Dorr and Hasan Altaf of the *Paris Review*; Austin Mueller of the Wylie Agency; and the Princeton University Press Permissions Department.

Fallout was, in part, written to honor the memory of my father, who brought me up in a broadcast newsroom and was a passionate advocate for ethical and neutral journalism. And this book might not exist at all if not for a question posited by my husband and longtime collaborator, Gregory Macek, that set the entire project in motion. As with my last book, *Everybody Behaves Badly*, *Fallout* belongs as much to him as it does to me. It honors our mutual newsroom roots and celebrates the things we hold most sacred, today more than ever: truth seeking, decency, and honor.

Notes

INTRODUCTION

1 *Not intended to write an exposé*: Hersey to Michael J. Yavenditti, as re-counted in "John Hersey and the American Conscience: The Reception of 'Hiroshima,'" *Pacific Historical Review* 43, no. 1 (February 1974): 24–49. Yavenditti corresponded with Hersey and interviewed him on September 19, 1967.

2 *More than 42,000 civilians*: "Statistics of Damages Caused by Atomic Bombardment, August 6, 1945," Foreign Affairs Section, Hiroshima City. As of August 25, 1945, the city of Hiroshima estimated that 21,135 civilian men and 21,277 civilian women had died, with 3,772 missing. By November 30, the city's estimate rose to 38,756 men dead and 37,065 women dead, with 2,329 missing. This was one of several damage and casualties resources upon which Hersey relied in writing "Hiroshima." John Hersey Papers, Beinecke Library, Yale University.

2 *Rise to 100,000*: See John Hersey, "Hiroshima," *New Yorker*, August 31, 1945, 15.

2   *As many as 280,000*: Hiroshima death estimates range widely, from a low of 68,000 (calculated by the U.S. Atomic Energy Commission) to a high of 280,000 (by *Chugoku Shimbun*, a Hiroshima newspaper). In 1970, Dr. Minoru Yuzaki, a sociologist and research fellow at the University of Hiroshima's Research Institute for Nuclear Medicine and Biology, conducted a casualty study and placed the death toll at around 200,000. For the study, Yuzaki attempted to re-create a house-by-house map of Hiroshima at the moment of impact. Yet even he called his estimate "tentative." For more information, see "Japan: To Count the Dead," *Time*, August 10, 1970.

2   *remains . . . uncovered today*: For example, in 1987, the bodies of sixty-four atomic bomb victims were uncovered at a popular Hiroshima park and garden called Asano Park by Hersey in "Hiroshima" and known as Shukkei-en Garden now. In a November 30, 2018, interview with Lesley Blume, Hiroshima Prefecture's governor, Hidehiko Yuzaki, stated that Hiroshima has never been wholly excavated.

2   *"You dig two"*: Governor Hidehiko Yuzaki interview with Lesley Blume, November 30, 2018.

2   *"expect a rain"*: President Harry S. Truman, "Statement by the President of the United States," White House Press Release, August 6, 1945, reproduced in full on the website of the Atomic Heritage Foundation: https://www.atomicheritage.org/key-documents/truman-statement-hiroshima.

3   *"clearly . . . those French," "So I changed,"* and *"my mistake became"*: Walter Cronkite, *A Reporter's Life* (New York: Alfred A. Knopf, Inc., 1996), 124.

3   *"stealing God's Stuff"*: E. B. White, *New Yorker*, August 18, 1945, 13.

3   *"What happened at"*: Sidney Shalett, "New Age Ushered; Day of Atomic Energy Hailed by President, Revealing Weapon," *New York Times*, August 7, 1945, 1.

3   *"most of us"*: Arthur Gelb, *City Room* (New York: G. P. Putnam's Sons, 2003), 103–4.

4 *"get the reputation"*: Henry L. Stimson, June 6, 1945, as quoted in Monica Braw, *The Atomic Bomb Suppressed: American Censorship in Japan* (Armonk, NY: M. E. Sharpe, Inc., 1991), 138.

4 *"disturb public tranquility"*: Office of the Supreme Commander for the Allied Powers Press Code, issued September 19, 1945, as reprinted in William Coughlin, *Conquered Press: The MacArthur Era in Japanese Journalism* (Palo Alto, CA: Pacific Books, 1952), 149–50, and Monica Braw, *The Atomic Bomb Suppressed*, 41.

4 *"It was just . . ."*: President Harry S. Truman, statement made at Columbia University, August 27–29, 1959, as quoted in Cyril Clemens, ed., *Truman Speaks* (New York: Columbia University Press, 1960), 93.

4 *"very pleasant way . . ."*: General Groves speaking to the Senate Special Committee on Atomic Energy, "Hearings: Atomic Energy Act of 1945," as quoted in Michael J. Yavenditti, "John Hersey and the American Conscience," *Pacific Historical Review* 43, no. 1 (February 1974), 27, and Sean Malloy, " 'A Very Pleasant Way to Die': Radiation Effects and the Decision to Use the Atomic Bomb against Japan," *Diplomatic History* 36 (June 2012).

5 *"spectacular; but . . . impersonal . . ."*: John Hersey, "The Mechanics of a Novel," *Yale University Library Gazette* 27, no. 1 (July 1952).

6 *deadliest conflict in human history*: World War II death toll estimates: National WWII Museum, referenced in January 2019: https://www.nationalww2museum.org/students-teachers/student-resources/research-starters/research-starters-worldwide-deaths-world-war.

6 *26.6 million Russians dead*: Statistic according to Major General Alexandr Kirilin, head of administration of the Ministry of Defense, as summarized in "The Ministry of Defense Clarifies Data on Those Killed in the Second World War," *Kommersant*, May 5, 2010, https://www.kommersant.ru/doc/1364563. *More than 407,000 Americans dead*: "U.S. Military Casualties in World War II," National WWII Museum, referenced in November 2019: https://www.nationalww2museum

.org/students-teachers/student-resources/research-starters/research
-starters-us-military-numbers.

7 *"When headlines say . . ."* and *"You swallowed statistics . . ."*: Lewis
Gannett, "Books and Things," *New York Herald Tribune*, August 29,
1946, 23.

8 *"American pride [had] . . ."*: John Hersey, "A Mistake of Terrifically
Horrible Proportions," *Manzanar* (New York: Times Books, 1988),
11–12.

8 *117,000 people of Japanese descent interned*: "Japanese Reloca-
tion During World War II," National Archives, referenced in November 2019: https://www.archives.gov/education/lessons/japanese
-relocation.

8 *"had been repaid . . ."*: President Harry S. Truman, "Statement by
the President of the United States," White House Press Release, August 6, 1945, reproduced in full on the website of the Atomic Heritage
Foundation: https://www.atomicheritage.org/key-documents/truman
-statement-hiroshima.

8 *85 percent approved*: Gallup poll, August 1945, cited in Robert Jay
Lifton and Greg Mitchell, *Hiroshima in America* (New York: G. P. Putnam's Sons, 1995), 33, and Michael J. Yavenditti, "John Hersey and
the American Conscience," 25. *23 percent wanted more bombs*: Roper
poll, Ibid, 33.

8 *"If our concept . . ."*: John Hersey, *Into the Valley* (New York: Schocken,
1989), xxx.

9 *"They still wondered . . ."*: John Hersey, "Hiroshima," 15.

10 *more than five hundred U.S. radio stations*: "A Survey of Radio Comment on the Hiroshima Issue of THE NEW YORKER, September 6,
1946, by Radio Reports, Inc.," *New Yorker* records, New York Public
Library.

11 *"escape into easy . . ."*: Albert Einstein, "The War Is Won, but the
Peace Is Not," speech, December 10, 1945, as reprinted in David E.
Rowe and Robert Schulmann, eds., *Einstein on Politics: His Private*

Thoughts and Public Stands on Nationalism, Zionism, War, Peace, and the Bomb (Princeton, NJ: Princeton University Press, 2007), 382.

11 *"has not been . . ."*: John Hersey, "John Hersey, The Art of Fiction No. 92," interview by Jonathan Dee, *Paris Review*, no. 100 (Summer–Fall 1986).

12 *Tsar Bomba*: "Tsar Bomba," Atomic Heritage Foundation, August 8, 2014, referenced November 25, 2019, https://www.atomic heritage.org/history/tsar-bomba.

12 *Approximately 13,000 warheads*: Matt Korda email to Lesley Blume, March 8, 2021.

12 *"I do not . . ."*: Albert Einstein interview with Alfred Werner, "Einstein at Seventy," *Liberal Judaism* (May–June 1949), as quoted in David E. Rowe and Robert Schulmann, eds., *Einstein on Politics: His Private Thoughts and Public Stands on Nationalism, Zionism, War, Peace, and the Bomb* (Princeton, NJ: Princeton University Press, 2007), 405.

12 *"slippage"*: John Hersey as quoted in "After Hiroshima: An Interview with John Hersey," *Antaeus Report*, Fall 1984, 4.

13 *"100 seconds to midnight"*: "Closer than ever: It is 100 seconds to midnight: 2020 Doomsday Clock Statement," *Bulletin of the Atomic Scientists*, John Mecklin, ed., January 23, 2020, https://thebulletin.org /doomsday-clock/current-time/.

13 *"the most dangerous . . ."* and *"The world is . . ."*: Dr. William J. Perry interviews with Lesley Blume on January 31, 2020, and February 5, 2019, and Dr. William J. Perry email to Lesley Blume on March 8, 2021.

13 *a third supported [North Korea] strike . . . , "It's our best . . . ,"* and *"to end North . . ."*: Alida R. Haworth, Scott Sagan, and Benjamin A. Valentino, "What do Americans Really Think about Conflict with Nuclear North Korea? The Answer is Both Reassuring and Disturbing," *Bulletin of the Atomic Scientists*, July 2, 2019: https://thebulletin.org/2019/07 /what-do-americans-really-think-about-conflict-with-nuclear-north -korea-the-answer-is-both-reassuring-and-disturbing/.

14 *"Big If"*: John Hersey, *Here to Stay* (New York: Alfred A. Knopf, Inc., 1963), 243.

CHAPTER ONE: The Picture Does Not Tell the Whole Story

15   *quarter of a million people*: Meyer Berger, "Lights Bring Out Victory Throngs," *New York Times*, May 9, 1945, 17.

15   *Over a thousand tons of paper*: The New York City Department of Sanitation announced that 1,074 tons of paper had been collected from the two-day celebration commemorating the unconditional German surrender: "Paper Salvage Lowered by V-E Day Celebrations," *New York Times*, May 10, 1945, 20.

15   *joyous, deafening cacophony*: "Life Goes to Some V-E Day Celebrations," *Life*, May 21, 1945, 118–21.

16   *"I just heard . . ."* and *"Lauterbach, you bastard . . ."*: John Hersey, "John Hersey, The Art of Fiction No. 92," interview by Jonathan Dee, *Paris Review*, issue 100 (Summer-Fall 1986).

16   *"I should have . . ."*: John Hersey, *Into the Valley* (New York: Schocken Books,1989), xxv.

17   *an invitation to the White House*: John Hersey Papers, Beinecke Rare Book & Manuscript Library, Yale University Library. Winchell mention of Hersey: "Winchell Coast-to-Coast," *Daily Mirror*, July 6, 1944; clipping also in John Hersey Papers at Beinecke.

17   *"most popular member"* and *"most influential"*: John McChesney, "John Hersey '32: The Novelist," *Hotchkiss Magazine*, July 1965.

17   *"hollow"* and *"flogging his wares"*: Baird Hersey as quoted in Russell Shorto, "John Hersey: The Writer Who Let 'Hiroshima' Speak for Itself," *New Yorker*, August 31, 2016. Baird also stated in this story that Hersey "wasn't a religious person—he eventually reacted against being raised in that world."

17   *"let his works . . ."*: Brook Hersey as quoted in Ibid.

17   *something of a cipher*: "John Hersey was not one to feel that his job was to sell himself," recalled his later editor at Knopf, Judith Jones, who added, "He didn't have an agent. He almost never gave interviews, and he shunned the idea of going out on the road to peddle his

wares." (Source: Judith Jones, VP, Knopf, "As Others Saw Him," *Yale Alumni Magazine*, October 1993.) That said, Hersey was dedicated to preserving his professional legacy from the earliest years of his career, gifting his "Hiroshima" materials to Yale University's Beinecke Rare Book & Manuscript Library and saving newspaper clippings, personal and professional letters, invitations, articles and profiles about himself, drafts and research materials for his works, photographs, and other autobiographical materials and memorabilia in a private collection. These personal and professional materials were also denoted to Yale, where they are now available to scholars; the self-created collection and donation indicates that Hersey knew his life and role as a public figure would be of interest and merit to future scholars, journalists, and biographers, and willingly furnished them with the materials to reconstruct his life, reporting, and creative processes.

18   *opened the* Time *bureau*: David Scott Sanders, unpublished notes ("John Hersey Interview, Expanded Notes") from an August 13, 1987, interview with John Hersey, 3.

18   *"to catch a . . ."*: John Hersey, "The Mechanics of a Novel," *Yale University Library Gazette* 27, no. 1 (July 1952).

18   *America, democracy, and free enterprise*: Luce's vision was neatly summarized in his statement that "the 20th century must be to a significant degree the American century." See the Luce biography by Alan Brinkley, *The Publisher: Henry Luce and His American Century* (New York: Random House, Inc., 2010) for more, and Alden Whitman, "Henry R. Luce, Creator of Time-Life Magazine Empire, Dies in Phoenix at 68," *New York Times*, March 1, 1967.

18   *"there was as . . ."*: Theodore H. White, *In Search of History: A Personal Adventure* (New York: Harper & Row, Publishers, Inc., 1978), 257.

19   *"walking wonder of . . ."*: John Hersey, "Henry Luce's China Dream," *New Republic*, May 2, 1983, as reprinted in John Hersey, *Life Sketches* (New York: Alfred A. Knopf, Inc., 1989), 27. *"Quasi-parental"*:

Notes

Thomas Griffith, *Harry & Teddy: The Turbulent Friendship of Press Lord Henry R. Luce and His Favorite Reporter, Theodore H. White* (New York: Random House, Inc., 1995), 141.

19 *Time's top job*: John Hersey, "Henry Luce's China Dream," as reprinted in John Hersey, *Life Sketches*, 38.

19 *Hersey resigned*: Robert E. Herzstein, *Henry R. Luce, Time, and the American Crusade in Asia*, 49.

19 *"On to Tokyo!"*: Meyer Berger, "Lights Bring Out Victory Throngs," *New York Times*, May 9, 1945, 17.

19 *"I had been . . ."*: John Hersey, "After Hiroshima: An Interview with John Hersey," *Antaeus Report*, Fall 1984, 3. Hersey stated that "it was thought that there would have to be an invasion of Japan and the prospective losses in such an invasion would have been horrendous on both sides." Ibid.

19 *begin diverting veterans*: Richard L. Strout, "V-E Day: A Grand Anticlimax for Some," *Christian Science Monitor*, May 8, 1945, 16.

20 *"bear of a . . ."*: Harrison E. Salisbury, *A Journey for Our Times: A Memoir* (New York: Harper & Row, Publishers, Inc., 1983), 251.

20 *"the G.I.s called . . ."*: Bill Lawrence, *Six Presidents, Too Many Wars* (New York: Saturday Review Press, 1972), 123–24.

20 *will of Japanese soldiers:* Bill Lawrence, *Six Presidents, Too Many Wars*, 125. He adds that, according to his reporting, intelligence officers had been estimating casualties of half a million Allied troops in a ground assault planned for fall 1945. Ibid., 126.

20 *"Few of us . . ."* : Ibid., 126.

20 *Cold Spring Harbor*: John Hersey to Michael Yavenditti, as depicted in Michael J. Yavenditti, "John Hersey and the American Conscience," *Pacific Historical Review* 43, no. 1 (February 1974), 35; and John Hersey to David Scott Sanders, as reflected in unpublished notes ("John Hersey Interview, Expanded Notes") from their August 13, 1987, interview, 5.

20 *"The force from . . ."*: President Harry S. Truman, "Statement by the

President of the United States," White House Press Release, August 6, 1945, reproduced in full on the website of the Atomic Heritage Foundation: https://www.atomicheritage.org/key-documents/truman-state ment-hiroshima.

21 *wasn't as bewildering*: John Hersey to David Scott Sanders, as reflected in unpublished notes ("John Hersey Interview, Expanded Notes") from their August 13, 1987, interview, 5.

21 *$2 billion nuclear undertaking*: President Harry S. Truman, "Statement by the President of the United States," White House Press Release, August 6, 1945, reproduced in full on the website of the Atomic Heritage Foundation: https://www.atomicheritage.org/key-documents /truman-statement-hiroshima. Also Jay Walz, "Atom Bombs Made in 3 Hidden 'Cities,'" *New York Times*, August 7, 1945, 1.

21 *"the slightest inkling . . ."* and *"All of them . . ."*: William L. Laurence, *Dawn Over Zero: The Story of the Atomic Bomb* (New York: Alfred A. Knopf, Inc., 1946), 197.

21 *sense of despair*: John Hersey to David Scott Sanders, as reflected in unpublished notes ("John Hersey Interview, Expanded Notes") from their August 13, 1987, interview, 5; John Hersey to Michael Yavenditti, as depicted in Michael J. Yavenditti, "John Hersey and the American Conscience," 35.

22 *a "totally criminal" action, "We gave the . . . ,"* and *"sure that one . . ."*: John Hersey, "After Hiroshima: An Interview with John Hersey," *Antaeus Report*, 3.

22 *"a terrifying factor . . ."*: Hersey to Michael J. Yavenditti, July 30, 1971, as quoted in Michael J. Yavenditti, "John Hersey and the American Conscience," 35.

22 *"living totem pole . . ."* and *"as though the . . ."*: William L. Laurence, "Atomic Bombing of Nagasaki Told by Flight Member," *New York Times*, September 9, 1945.

22 *"An impenetrable cloud . . ."* and *"what happened at . . ."*: Sidney Shalett, "First Atomic Bomb Dropped on Japan; Missile Is Equal to

20,000 Tons of TNT; Truman Warns Foe of a 'Rain of Ruin,' " *New York Times*, August 7, 1945, 1.

22 *downplay the attacks*: Monica Braw, *The Atomic Bomb Suppressed: American Censorship in Japan* (Armonk, NY: M. E. Sharpe, Inc., 1991), 11. At first, the Japanese government and mass media suppressed the true magnitude of the attack—and the fact that the weapon used was an atomic bomb, as opposed to a conventional weapon—even though the Japanese government had been fully briefed by August 8 by Japan's leading nuclear physicist, Professor Yoshio Nishina, who had been dispatched immediately to Hiroshima to survey the damage. In his report, Professor Nishina described the scene as "unspeakable," with tens of thousands dead: "Bodies piled up everywhere . . . sick, wounded, naked people wandering around in a daze . . . almost no buildings left standing." He regretted having to inform the government that the new weapon that had been used was an atomic bomb (Ibid., 12). He had been able to identify it as such because Japan, like Germany, had been conducting its own atomic research, although the Manhattan Project's General Leslie Groves had "been quite certain that Japan posed no danger" when it came to the possibility of competition. (Source: Robert S. Norris, *Racing for the Bomb* [South Royalton, VT: Steerforth Press L.C., 2002], 450.)

22 *"Hiroshima was attacked . . ."*: *Asahi Shimbun* article quoted in Monica Braw, *The Atomic Bomb Suppressed*, 11.

23 *comprehended their situation*: Ibid., 14.

23 *"parachute-borne atomic bombs"* and *" by employing a . . ."*: "Tokyo Radio Says Hiroshima Hit by Parachute Atomic Bombs," United Press, August 7, 1945.

23 *"new and most . . ."* and *"to [the] total . . ."*: Emperor Hirohito's August 15, 1945, surrender radio address, as quoted in Monica Braw, *The Atomic Bomb Suppressed*, 17.

23 *"the victory roar . . . ,"* *"wild and instantaneous,"* and *"metropolis exploded its . . ."*: Alexander Feinberg, "All City 'Lets Go': Hundreds of

Thousands Roar Joy After Victory Flash is Received," *New York Times*, August 15, 1945, 1.

24 *"over-exuberance"*: "City Police Prepared for V-J Celebration," *New York Herald Tribune*, August 9, 1945, 2B.

24 *"HANG THE EMPEROR"*: Alexander Feinberg, "All City 'Lets Go,'" 1.

24 *vast majority . . . approved*: 85 percent of Americans approved of use of the bomb, as per a Gallup poll taken in August 1945, as cited in Robert Jay Lifton and Greg Mitchell, *Hiroshima in America* (New York: G. P. Putnam's Sons, 1995), 33, and Michael J. Yavenditti, "John Hersey and the American Conscience," 25. *Even more atomic bombs:* 23 percent of Americans polled in an August 1945 poll conducted by Roper said that they regretted that more atomic bombs had not been used pre-surrender: Robert Jay Lifton and Greg Mitchell, *Hiroshima in America*, 33.

24 *"defeated and destroyed . . ."* and *"within an hour . . ."*: Fiorello La Guardia, August 15, 1945, radio address, as quoted in "Mayor Proclaims Two Victory Days," *New York Times*, August 15, 1945, 6.

25–26 *"continu[e] to die . . . ," "the majority of . . . ," "inhalation of the . . . ,"* and *"inhaled uranium"*: Leslie Nakashima, "Hiroshima as I Saw it," United Press, August 27, 1945.

26 *"United States scientists . . ."*: "Hiroshima Gone, Newsman Finds," *New York Times*, August 31, 1945, 4.

26 *"Japanese reports of . . ."* and *"I think our . . ."*: "Japanese Reports Doubted," *New York Times*, August 31, 1945, 4.

26–27 *"warning to the world . . . ," "mysteriously and horribly," "plague," "it is given . . ."* and *"THE PICTURE THAT . . . "*: Wilfred Burchett, "The Atomic Plague," *Daily Express*, September 5, 1945, 1. Text also reproduced in Wilfred Burchett, *Shadows of Hiroshima* (London: Verso Editions, 1983), 34–36.

27 *"the fate of . . ."*: *Public Enemy Number One* (documentary about Wilfred Burchett, directed by David Bradbury), 1981.

27 *"never looked upon . . . ," "the awful, sickening . . . ,"* the bomb's lingering . . . , and *"vomited blood and . . ."*: W. H. Lawrence, "Visit to

Hiroshima Proves It World's Most-Damaged City," *New York Times*, September 5, 1945, 1.

27 *"FOE SEEKS TO . . ."* and *"horrible as the . . ."*: W. H. Lawrence, "Atom Bomb Killed Nagasaki Captives," *New York Times*, September 9, 1945.

28 *uneasy and upset*: Michael Yavenditti, "American Reactions to the Use of Atomic Bombs on Japan, 1945–1947," dissertation for doctorate of philosophy, University of California, Berkeley, 1970, 362.

28 *"Most of it . . ."*: Bill Lawrence letter to John Hersey, September 10, 1945, John Hersey Papers, Beinecke Library, Yale University.

29 *Wildcat Farm*: Charles J. Kelly, *Tex McCrary: Wars, Women, Politics: An Adventurous Life Across the American Century* (Lanham, MD: Hamilton Books, 2009), 1.

29 *pioneer the morning talk show*: Richard Severo, "Tex McCrary Dies at 92; Public Relations Man Who Helped Create Talk-Show Format," *New York Times*, July 30, 2003, C12.

29 *"What, another?"* and *"earth-shaking event which . . ."*: Clark Lee of INS, as quoted in Dickson Hartwell and Andrew A. Rooney, *Off the Record: Inside Stories from Far and Wide Gathered by Members of the Overseas Press Club* (New York: Doubleday & Company, Inc., 1953), 225. McCrary biographer Charles J. Kelly stated on page 79 of *Tex McCrary* that the original purpose of the junket was "to do whatever he could to make sure the Air Force's role was seen by the American people as the strategic weapon in the defeat of the enemy," but several reporters selected for the junket recall specifically being told from the beginning that they had been tapped to report on the atomic bombings as the culminating major event of the war.

29 *His request was declined*: Charles J. Kelly, *Tex McCrary*, 89.

29 *two gleaming Boeing B-17s*: Bill Lawrence to John Hersey, September 10, 1945, and Clark Lee of INS, as quoted in Dickson Hartwell and Andrew A. Rooney, *Off the Record*, 225.

29 *"compare it with . . ."*: Clark Lee of INS, as quoted in Dickson Hartwell and Andrew A. Rooney, *Off the Record*, 226.

29–30 *got the news about Hiroshima:* Clark Lee and other members of the junket later stated that they had been slated to witness the actual bombing of Hiroshima but that their junket had been held up when McCrary diverted the planes to Rome for a visit with his wife, actress and model Jinx Falkenburg, who was in the Italian city doing a USO show. When Falkenburg gave birth to the couple's baby nine months later, the timing implied that it was conceived during the Rome visit; one noted dryly that the baby was nicknamed "Hero"—short for Hiroshima—to commemorate "the time when love proved stronger than the atomic bomb." Clark Lee as quoted in Dickson Hartwell and Andrew A. Rooney, *Off the Record*, 227.

30 *"I ad-libbed my . . .":* Bill Lawrence, *Six Presidents, Too Many Wars*, 131–32.

30 *"death laboratory"* and *"human guinea pigs":* Clark Lee, *One Last Look Around* (New York: Duell, Sloan, and Pearce, 1947), 77.

30 *"housetrained reporters":* Wilfred Burchett, *Shadows of Hiroshima*, 17. *"Being rewarded for . . .":* Ibid., 15.

31 *"not ready for . . .":* Charles J. Kelly, *Tex McCrary*, 98.

31 *court-martial the entire entourage:* Bill Lawrence, *Six Presidents, Too Many Wars*, 139.

31 *Morse code handset:* Wilfred Burchett, *Shadows of Hiroshima*, 21, 42–43.

31 *"press ghetto"* : Wilfred Burchett, *Shadows of Hiroshima*, 24.

31 *sentries on the bridges:* Charles J. Kelly, *Tex McCrary*, 96.

31 *"General MacArthur had . . ."* : Wilfred Burchett, *Shadows of Hiroshima*, 23.

32 *"much machine-gunned George . . .":* George Weller, *First into Nagasaki: The Censored Eyewitness Dispatches on Post-Atomic Japan and Its Prisoners of War*, edited by Anthony Weller (New York: Three Rivers Press, 2006), 251.

32 *"I had a . . .":* George Weller, *First into Nagasaki*, 4–5; *"a long cold . . .":* Ibid., 276.

32 *"like yacht passengers . . ."*: George Weller, *First into Nagasaki*, 19.

33 *"party,"* *"fabulous trip,"* and *"I have been . . ."*: Bill Lawrence letter to John Hersey, September 10, 1945, John Hersey Papers, Beinecke Library, Yale University.

33 *"As a journalist . . ."*: John Hersey, "After Hiroshima: An Interview with John Hersey," *Antaeus Report*, Fall 1984, 2.

CHAPTER TWO: Scoop the World

35 New York Times *headquarters . . . original building*: "From Dazzling to Dirty and Back Again: A Brief History of Times Square," Times Square: The Official Website: https://www.timessquarenyc.org/history-of-times-square.

35 *"chateauesque"* and *"the greatest and . . ."*: "News Paper Spires: From Park Row to Times Square: New York Times Annex," Skyscraper Museum website: https://www.skyscraper.org/EXHIBITIONS/PAPER_SPIRES/nw14_ta.php.

35 *"Squalor had come . . ."*: Brendan Gill, *Here at* The New Yorker (New York: Random House, Inc., 1974), 104. *"Bleak little ill-painted . . ."*: Ibid., 4–5.

36 *"Too many people . . ."*: Harold Ross as quoted in Brendan Gill, *Here at* The New Yorker, 370.

36 *"the old lady . . ."*: Harold Ross, "The New Yorker Prospectus," Fall 1924, as reprinted in Thomas Kunkel, *Genius in Disguise* (New York: Carroll & Graf Publishers, Inc., 1995), 439–40.

36 *"Backward ran . . ."*: Wolcott Gibbs, "Time . . . Fortune . . . Life . . . Luce," *New Yorker*, November 28, 1936.

37 *chic Manhattan nightclub*: Regarding the nightclub setting of the JFK meeting: Hersey later could not recall which club it was. (Source: "John Hersey, Interview with Herbert Farmet," Oral History Research Office, Columbia University, December 8, 1976, 8–9.) Kennedy biographer Michael O'Brien states that the meeting took place on February 9,

1944 at the Stork Club. (Source: Michael O'Brien, *John F. Kennedy: A Biography* [New York: St. Martin's Press, 2005], 170.) Ben Yagoda's *New Yorker* biography, *Around Town* (New York: Scribner, 2000), 184, states that the nightclub was La Martinique.

37 *"He was then . . ."* and *"it was a . . ."*: "John Hersey, interview with Herbert Farmet," Oral History Research Office, Columbia University, December 8, 1976, 8–9.

37 *brought it to William Shawn*: Hersey may have been first introduced to the *New Yorker* team by his Yale classmate Brendan Gill—*New Yorker* writer and eventual biographer of the magazine—according to Gill's son, Michael Gill Gates. Brendan Gill had been trying to lure Hersey over to the *New Yorker* and away from *Time* magazine, which Gill considered to be a "port of last resort" publication. Michael Gates Gill interview with Lesley Blume, January 29, 2018.

37 *"been trying for . . ."* and *"in a glow . . ."*: Harold Ross to Joseph Kennedy, May 18, 1944, *New Yorker* records, New York Public Library.

38 *"Up to [his] . . ."*: Harold Ross letter to Alexander Woollcott, May 19, 1942, as excerpted in Thomas Kunkel, *Genius in Disguise*, 354.

38 *"He had a . . ."* and *"the editor of . . ."*: John Hersey, "Note," *Life Sketches* (New York: Alfred A. Knopf, 1989), ix.

38 *brandished a knitting needle*: Janet Flanner, "Introduction: The Unique Ross," in Jane Grant, *Ross, The New Yorker, and Me* (New York: Reynal and Company, Inc., 1968), 12. *"All right . . . God bless you"*: E. B. White obituary of Harold Ross, as quoted in Brendan Gill, *Here at The New Yorker*, 297.

38 *"being stung to . . ."*: Brendan Gill, *Here at* The New Yorker, 7.

38 *"Interrogatory howls of . . ."*: John Hersey, "Note," *Life Sketches*, x.

38 *"I am there . . ."*: Lillian Ross, *Here but Not Here: A Love Story* (New York: Random House, Inc., 1998), 11.

38 *resembled an elf*: Michael Gates Gill email to Lesley Blume, January 31, 2018.

38–39 *"thirty-seven, flat-footed, stoop-shouldered, . . ."*: Harold Ross letter to Stephen T. Early, March 14, 1944, as quoted in Ben Yagoda, *About Town*, 182.

39 *"He was extraordinarily . . ."*: John Bennet interview with Lesley Blume, February 7, 2018.

39 *"every human being . . ."*: Lillian Ross, *Here but Not Here*, 15. *"Even Hitler?"*: Ibid., 50.

39 *"we don't cover . . ."*: William Shawn, untitled essay, in Brendan Gill, *Here at* The New Yorker, 392.

39 *"world of speakeasies . . ."*: Brendan Gill, *Here at* The New Yorker, 88.

39 *"Nobody feels funny . . ."*: Harold Ross quoted in Jane Grant, *Ross, The New Yorker, and Me*, 261.

39 *"the greatest journalistic . . ."*: William Shawn to John Bennet, as recounted in Bennet interview with Lesley Blume, February 7, 2018.

40 *"hunch man"*: Harold Ross nicknamed Shawn the "hunch man," according to Thomas Kunkel, *Genius in Disguise*, 352.

40 *"pulverised by bombs"*: Janet Flanner, "Letter from Cologne," *New Yorker*, March 31, 1945

40 *"How simple it . . ."*: Harold Ross letter to Janet Flanner, March 27, 1945, as excerpted in Ben Yagoda, *About Town*, 180.

40 *"any number of . . ."* and *"unless the* New *. . ."*: Harold Ross letter to Janet Flanner, March 27, 1945, as excerpted in Ben Yagoda, *About Town*, 181.

40 *"We published what . . ."*: William Shawn, untitled essay in Brendan Gill, *Here at* The New Yorker, 391.

40 *with only minor changes*: Hersey's Kennedy profile for the *New Yorker* was submitted to the Navy's Office of Public Relations, whose representatives returned it with "no objections to its publication." Letter from Allan R. Jackson, LT. USNR, Navy Dep Office of Public Relations to William Shawn, May 25, 1944, *New Yorker* records, New York Public Library.

41 *"very important experience"*: "John Hersey, Interview with Herbert

Farmet," Oral History Research Office, Columbia University, December 8, 1976, 10.

41 *"had [a] genius . . ."*: John Hersey to David Scott Sanders, as reflected in unpublished notes ("John Hersey Interview, Expanded Notes") from their August 13, 1987, interview, 5.

41 *"We have long . . ."* and *"we are not . . ."*: Harold Ross to Ambassador Joseph Kennedy, May 18, 1944, *New Yorker* records, New York Public Library, Box 52.

41 *100,000 copies of "Survival"*: Robert Dallek, *An Unfinished Life: John F. Kennedy, 1917–1963* (New York: Little, Brown, and Company, 2003), 130, and Michael O'Brien, *John F. Kennedy*, 201–2.

42 *"[Shawn's] idea of a . . ."*: Eric Pace, "William Shawn, 85, is Dead; *New Yorker*'s Gentle Despot," *New York Times*, December 9, 1992, A1.

42 *couple of military-friendly pieces*: Hersey's 1945 stories for the *New Yorker* include "The Brilliant Jughead," July 20, 1945, and "Long Haul, with Variables," August 31, 1945.

42 *boast to advertisers*: "The Biggest Story in History. . . and He Was the Only Reporter There!," *New York Times* circular to advertisers, undated, *New York Times* archives. *"biggest news story. . ."*: Robert Simpson, "The Infinitesimal and the Infinite," *New Yorker*, August 18, 1945, 26.

42 *"Most of the . . ."*: John Hersey, "After Hiroshima: An Interview with John Hersey," *Antaeus Report*, Fall 1984, 2.

43 *A spread of Eyerman's photos*: "The Tokyo Express: A *Life* Photographer Takes a Ride to Hiroshima on Japan's Best Train," *Life*, October 8, 1945, 27–35.

44 *Rather, the War Department*: September 14, 1945, War Department press release to publication editors, paraphrased in Monica Braw, *The Atomic Bomb Suppressed* (Armonk, NY: M. E. Sharpe, Inc., 1991), 111.

44 *"the highest national . . ."*: September 14, 1945, War Department press

release to publication editors, as quoted in George Weller, *First into Nagasaki: The Censored Eyewitness Dispatches on Post-Atomic Japan and Its Prisoners of War*, edited by Anthony Weller (New York: Three Rivers Press, 2006), 266.

44   *"This might be . . ."*: Charles Ross memo to Lieutenant Colonel B. W. Davenport, August 27, 1945, as quoted in Robert Jay Lifton and Greg Mitchell, *Hiroshima in America* (New York: G. P. Putnam's Sons, 1995), 51.

44–45 *"on loan from . . ."*: "William Laurence: Science Reporter," Internal *New York Times* employee sketch, June 20, 1952, *New York Times* archives. *"make certain that . . . ," "the Japanese are . . . ," "the Japanese claim," "gave mute testimony . . . ," "Tokyo tales . . . ,"* and *"radiations on the . . ."*: William L. Laurence, "U.S. Atom Bomb Site Belies Tokyo Tales," *New York Times*, September 12, 1945, 1.

45   *"simply used a . . ."*: Dr. Philip Morrison, quoted in Daniel Lang, "A Fine Moral Point," *New Yorker*, June 8, 1946, 69.

45–46 *"spot check," "The bomb had . . . ,"* and *"there had been . . ."*: Dr. Philip Morrison, quoted in Daniel Lang, "A Fine Moral Point," *New Yorker*, June 8, 1946, 62 and 76.

46   *indignant cables from afar*: Dr. Philip Morrison, quoted in Daniel Lang, "A Fine Moral Point," *New Yorker*, June 8, 1946, 62.

46   *"You could live . . ."*: General Leslie Groves as quoted in William L. Laurence, "U.S. Atom Bomb Site Belies Tokyo Tales," *New York Times*, September 12, 1945, 1.

46   *had been exaggerated*: William H. Lawrence, "No Radioactivity in Hiroshima Ruin," *New York Times*, September 13, 1945, and Robert Jay Lifton and Greg Mitchell, *Hiroshima in America*, 54.

46   *"those I had . . ."* and *"I'm afraid you've . . ."*: Wilfred Burchett, *Shadows of Hiroshima* (London: Verso Editions, 1983), 22–23.

46   *"No Radioactivity in . . ."* and *68,000 buildings destroyed*: W. H. Lawrence, "No Radioactivity in Hiroshima Ruin," *New York Times*, September 13, 1945, 4.

47 *"is a very . . ."*: General Leslie Groves, as quoted in Michael J. Yavenditti, "John Hersey and the American Conscience," *Pacific Historical Review* 43, no. 1 (February 1974), 27, and Sean Malloy, " 'A Very Pleasant Way to Die': Radiation Effects and the Decision to Use the Atomic Bomb Against Japan," *Diplomatic History* 36 (June 2012).

47 *"not an inhuman . . ."*: General Leslie Groves, Comments at Oak Ridge, August 29, 1945, as quoted in Robert S. Norris, *Racing for the Bomb* (South Royalton, VT: Steerforth Press L.C., 2002), 441.

47 *clearly had not sunk in*: Harold Ross and William Shawn would include a short editors' note on the first page of Hersey's "Hiroshima" in the August 31, 1945, issue of the *New Yorker*, stating that they were publishing the story, in part, because of their "conviction that few of us have yet comprehended the all but incredible destructive power of this weapon and that everyone might well take time to consider the terrible implications of its use." (Source: "To Our Readers," *New Yorker*, August 31, 1945, 15.)

47 *"what happened not . . ."*: John Hersey, "After Hiroshima: An Interview with John Hersey," *Antaeus Report*, Fall 1984, 2.

48 *story on the decimation of Cologne*: Ben Yagoda, *About Town*, 185.

48 *"wanton savagery"*: John Hersey, *Into the Valley*, xxix. *an animalistic enemy*: In one 1942 story, he refers to the Japanese as "our animal adversary." (Source: John Hersey, "The Marines on Guadalcanal," *Life*, November 9, 1942.)

48 *"I had long . . ."* and *"But I did . . ."*: John Hersey, *Into the Valley* (New York: Schocken, 1989), 49 and 67.

48 *"depravity of man"*: John Hersey, "The Mechanics of a Novel," *Yale University Library Gazette* 27, no. 1 (July 1952).

48 *civilization was to . . ."*: John Hersey, *Into the Valley*, xxx.

49 *pretended to be*: Wilfred Burchett, *Shadows of Hiroshima*, 28–30.

50 *"Of the fact . . ."* and *"determined to do . . ."*: George Weller, *First into Nagasaki*, 276–77.

50   *kept close tabs*: Surviving SCAP PRO Tokyo files at the U.S. National Archives and Records Administration (NARA) contain ample records detailing the tracking of reporters and their activities. From April through June 1946 alone, a sampling of surviving SCAP documents include reports such as "Report on Press, Speeches, Publications, and Motion Pictures (May 1, 1946)," "Report on Speech, Press, and Motion Pictures (May 15, 1946)," "Report on Speech, Publications, and Motion Pictures (May 30, 1946)." Notably, there was one May 1946 report titled "Activities of Time-Life International in Japan." Any reports on Hiroshima had to be submitted for consideration for "Permission to Publish." (Source: NARA, SCAP, "List of Papers, No. 000.76, File #3, Sheet #1, April 23—June 24, 1946.) Even more may have existed at one point; one seasoned NARA archivist estimates that only 1 to 3 percent of SCAP records have survived.

50   *Occupation troops*: Greg Mitchell, *Atomic Cover-up: Two U.S. Soldiers, Hiroshima & Nagasaki, and the Greatest Movie Never Made* (New York: Sinclair Books, 2012), 30–31. "Marines from the 2nd Division, with three regimental combat teams, took Nagasaki while the U.S. Army's 24th and 41st divisions seized Hiroshima." Ibid., 30.

50   *set up camp*: Greg Mitchell, *Atomic Cover-up*, 30.

51   *"downright quackery and . . ."*: General Douglas MacArthur cable to Warcos (PRO), November 2, 1946, SCAP Papers, NARA.

51   *"iron curtain of . . ."*: George Weller, "The Iron Curtain of Censorship," attempted dispatch to the *Chicago Daily News*, August 22, 1945, as excerpted in Anthony Weller, ed., *Weller's War* (New York: Three Rivers Press, 2009), 606.

51   *approved by the War Department*: When Hersey signed the contract with Knopf for *Men on Bataan*, the publisher headlined its March 12, 1942, newsletter with the news and also informed readers that "Mr. Hersey is writing it with the assistance and approval of the War Department."

51 *out of print*: John Hersey, "John Hersey: The Art of Fiction No. 92," *Paris Review*, interview by Jonathan Dee, issue 100, Summer–Fall 1986. *"too adulatory"*: John Hersey to David Scott Sanders, August 13, 1987, as quoted in unpublished notes ("John Hersey Interview, Expanded Notes") from their August 13, 1987, interview, 3.

51 *"a cruel and . . . ," "lost his marbles,"* and *"Seemed to me . . ."*: John Hersey, American Audio Prose Library interview with Kay Bonetti, 1988.

CHAPTER THREE: MacArthur's Closed Kingdom

53 *"animal adversary"*: John Hersey, "The Marines on Guadalcanal," *Life*, November 9, 1942. *"Stunted physically"*: John Hersey, "Joe Grew, Ambassador to Japan: America's Top Career Diplomat Knows How to Appease the Japanese or Be Stern with Them," *Life*, July 15, 1940.

53 *some helmet netting*: "The Battle of the River," *Life*, November 23, 1942, 99.

53 *"a swarm of . . ."*: John Hersey, *Into the Valley* (New York: Schocken Books, Inc., 1989), 65.

53 *"maybe the Japanese . . ."*: Russell Brines, *MacArthur's Japan* (Philadelphia: J. B. Lippincott Company, 1948), 23.

54 *"difference between Japanese . . ."*: General Leslie Groves in conversation with Lieutenant Charles Rea, doctor at Oak Ridge Hospital, as quoted in Robert Jay Lifton and Greg Mitchell, *Hiroshima in America* (New York: G. P. Putnam's Sons, 1995), 45.

54 *"Like other Americans . . ."* and *"horrified . . . by accounts . . ."*: John Hersey, *Into the Valley*, xxviii.

54 *"I had seen . . . ,"*: John Hersey, *Into the Valley*, xxviii–xxix.

54 *relentless bombardment*: John Hersey, Interview with Kay Bonetti, Hersey American Audio Prose Library, 1988.

54 *"started huge, uncontrollable . . ."*: John Hersey, "Letter from Chung-king," *New Yorker*, March 7, 1946.

54 *"You are doing . . ."*: William Shawn cable to John Hersey, March 1, 1946, *New Yorker* records, New York Public Library.

55 *"We had a . . ."* and *representatives of nations*: Bob Considine, *It's All News to Me: A Reporter's Disposition* (New York: Meredith Press, 1967), 199.

55 *"The more time . . ."*: William Shawn cable to John Hersey, March 22, 1946, *New Yorker* records, New York Public Library.

55 *successfully exploded*: "Operation Crossroads: Fact Sheet," Naval History and Heritage Command, U.S. Navy official website: https://www.history.navy.mil/about-us.html.

55 *"mirthless jesting about . . ."*: Bob Considine, *It's All News to Me*, 199.

56 *"I was just . . ."*: Norman Cousins recollection quoted in Robert Jay Lifton and Greg Mitchell, *Hiroshima in America*, 84.

56 *"the sound of . . ."*: Bob Considine, *It's All News to Me*, 202.

56 *no chilled spines*: Clark Lee, *One Last Look Around* (New York: Duell, Sloan and Pearce, 1947), 293–94.

56 *"They underwrote the . . ."*: Clark Lee, *One Last Look Around*, 294.

56 *"small, dark man"*: Robert Simpson, "The Infinitesimal and the Infinite," *New Yorker*, August 18, 1945, 28.

56 *a "foe's propaganda . . ."*: William L. Laurence, "U.S. Atom Bomb Site Belies Tokyo Tales; Tests on New Mexico Range Confirm That Blast, and Not Radiation, Took Toll," *New York Times*, September 12, 1945, 1.

56–57 *"greatest explosion ever . . ."* and *"churning mass of . . ."*: William L. Laurence, "Blast Biggest Yet," *New York Times*, July 25, 1946.

57 *less damage than expected*: According to an August 30, 1946, Gallup poll, as reported in "American Reactions to the Use of Atomic Bombs on Japan, 1945–1947" by Michael Yavenditti, dissertation for doctorate of philosophy, University of California, Berkeley, 1970, 333.

57 *moved military personnel*: Greg Mitchell, *Atomic Cover-up: Two U.S.*

Soldiers, Hiroshima & Nagasaki, and the Greatest Movie Never Made (New York: Sinclair Books, 2012), 33.

57 *"helping to give . . ."*: John Hersey, "A Reporter in China: Two Weeks' Water Away—II," *New Yorker*, May 25, 1946.

57 *transferred to a destroyer*: John Hersey, "John Hersey: The Art of Fiction No. 92," interview by Jonathan Dee, *Paris Review*, issue 100, Summer–Fall 1986.

58 *"terribly complicated story"*: John Hersey, "After Hiroshima: An Interview with John Hersey," *Antaeus Report*, Fall 1984, 2.

58 *"moment of shared . . ."*: John Hersey, "John Hersey: The Art of Fiction No. 92."

58 *"when the earth . . ."* and *"a giant tree . . ."*: William L. Laurence, "Blast Biggest Yet," *New York Times*, July 25, 1946, 2.

58 *"The human mind . . ."*: William L. Laurence, *Dawn Over Zero: The Story of the Atomic Bomb* (New York: Alfred A. Knopf, 1946), 224.

59 *"My hope was . . ."*: John Hersey, American Audio Prose Library Interview with Kay Bonetti, 1988.

59 "No objection on . . .": Cable from 1st Lt. Cav. Robert F. Jobson to John Hersey, May 13, 1946, John Hersey Papers, Beinecke Library, Yale University.

59 *"hereby invited and . . ."* : Cable from Lt. General Gillem to John Hersey, May 21, 1946, John Hersey Papers, Beinecke Library, Yale University.

60 *"The incendiaries dropped . . ."*: Bill Lawrence, *Six Presidents, Too Many Wars* (New York: Saturday Review Press, 1972), 121.

60 *"In Tokyo streets . . ."*: John Hersey, "Joe Grew, Ambassador to Japan: America's Top Career Diplomat Knows How to Appease the Japanese or Be Stern with Them," *Life*, July 15, 1940.

60 *"an ashtray filled . . ."*: George Weller, *Weller's War*, edited by Anthony Weller, 602.

60–61 *"occupationaires"* and *"flowed with unrelieved . . ."*: Russell Brines, *MacArthur's Japan*, 291.

61 *"and so many . . ."*: Ibid., 71.

61 *a modern Caesar*: Ray C. Anderson, Ph.D, M.D., *A Sojourn in the Land of the Rising Sun: Japan, the Japanese, and the Atomic Bomb Casualty Commission: My Diary, 1947–1949* (Sun City, AZ: Elan Press, 2005), 443.

61 *"When I walk . . ."* and *"The Japs are . . ."*: Ray C. Anderson, Ph.D, M.D., *A Sojourn in the Land of the Rising Sun*, 15.

61 *"one of the . . ."*: Dr. Masakazu Fujii, as quoted in Norman Cousins, "John Hersey," *Book-of-the-Month Club News*, March 1950.

61 *"one of America's . . . ," "tall, handsome [and] . . . ,"* and *"passing through Tokyo . . ."*: *Nippon Times*, "Over Here," September 22, 1946.

62 *an Army filmmaker*: Greg Mitchell, *Atomic Cover-up*, 22–23.

62 *"We knew that . . ."*: Ibid., 23.

62 *his hometown*: Ibid., 24.

62 *90,000 feet of footage*: Greg Mitchell, "The Great Hiroshima Cover-Up—And the Greatest Movie Never Made," *Asia-Pacific Journal* 9, issue 31, no. 4 (August 8, 2011).

63 *home to nearly every Western journalist*: Charles Pomeroy, ed., *Foreign Correspondents in Japan: Reporting a Half Century of Upheavals; From 1945 to the Present* (Rutland, VT: Charles E. Tuttle Company, 1998), 14. It is unclear whether Hersey lived on-site at the club during his brief stay in Tokyo before going to Hiroshima. Club historian Charles Pomeroy states that the archivist at the club—which still exists, albeit in a different location—could not find records of Hersey staying there but says that Hersey "no doubt visited and most likely stayed here." (Source: Charles Pomeroy email to Lesley Blume, May 26, 2018.) He added that the current club researcher searched surviving club records and that "no records of John Hersey's name was found in the FCCJ [i.e., Foreign Correspondents Club of Japan, as the Club is now called] archives of 1946," but that "only a few key records remain from that time." (Source: Charles Pomeroy email to Lesley Blume, May 28, 2018.)

63 *"a makeshift bordello . . ."*: Richard Hughes, as quoted in Charles Pomeroy, ed., *Foreign Correspondents in Japan*, 16.

63 *"writer from New . . ."*: Leslie Sussan, untitled, unpublished biography of her father, Herbert Sussan. Hersey may have been referred to Lieutenant Colonel Daniel McGovern by one of the other occupation journalists or by Tokyo-based Time-Life reporter John Luten to whom Hersey directed William Shawn to send his correspondence. In any case, a note about Lieutenant Colonel McGovern, his office location, and the fact that he made films about Hiroshima is cited in Hersey's surviving Japan notes at Yale's Beinecke Library.

63 *"limited strictly to . . ."*: General Orvil Anderson to Herbert Sussan, as quoted in Greg Mitchell, *Atomic Cover-up*, 25.

63 *"didn't want that . . ."*: McGovern quoted in Robert Jay Lifton and Greg Mitchell, *Hiroshima in America*, 59.

63 *"If people could . . ."*: Sussan quoted in Greg Mitchell, *Atomic Cover-up*, 23.

64 *footage of atomic bomb victims*: Greg Mitchell, *Atomic Cover-up*, 23–24.

64 *some Hiroshima contacts* and details of the meeting: Leslie Sussan, untitled, unpublished biography of her father, Herbert Sussan; Greg Mitchell, "The Great Hiroshima Cover-Up—And the Greatest Movie Never Made"; and Robert Jay Lifton and Greg Mitchell, *Hiroshima in America*, 259. Now declassified, 1946 footage of the Hiroshima priests filmed by the McGovern film unit can be seen online.

64 *Pacific pony edition*: A copy of the February 11, 1946, Pacific pony edition of *Time* containing the Siemes testimony is among Hersey's personal papers at Yale's Beinecke Library. A longer version of the Siemes report from the USSBS Restricted files (Memorandum, "Eyewitness Account of the Bombing of Hiroshima," Headquarters, U.S. Strategic Bombing Survey [Pacific], December 6, 1945) is also in his papers, filed among his primary research materials for "Hiroshima." Hersey later stated to a *Library Journal* reviewer that he had read the Siemes report before going to Hiroshima, which was what led him

to the Jesuit priests in that city. (Source: Letter from John Hersey to Robert H. Donahugh, July 21, 1985, John Hersey Papers, Beinecke Library, Yale University.)

64 *"The whole valley . . ."*: Account of "Rev. John A. Siemes, S.J., professor of modern philosophy at Tokyo's Catholic University," *Time* Pacific pony edition, February 11, 1946, 10.

64 *"our rescuing angel"*: Ibid.

65 *wartime camouflage*: Mark Gayn, *Japan Diary* (Rutland, VT: Charles E. Tuttle Company, 1981), 3. *reeked of fish*: Joseph Julian, *This Was Radio: A Personal Memoir* (New York: The Viking Press, 1975), 132.

65 *a large newsroom*: Charles Pomeroy, ed., *Foreign Correspondents in Japan*, 13. A note about Radio Tokyo and its second floor as a place from which to send cables are included in Hersey's surviving Japan notes, John Hersey Papers, Beinecke Library, Yale University.

65 *write the story back in the States*: Cable from William Shawn to John Hersey, March 22, 1946, *New Yorker* records, New York Public Library.

65 *"a tight grip . . ."* and *"They were told . . ."*: Russell Brines, *MacArthur's Japan*, 293.

66 *"Call it whimsy . . ."*: General Legrande Diller as quoted in William J. Coughlin, *Conquered Press: The MacArthur Era in Japanese Journalism* (Palo Alto, CA: Pacific Books, 1952), 116. ("Killer Diller" nickname: Ibid., 111.)

66 *"going to get . . ."* and *"Don't forget the . . ."*: General Legrande Diller as quoted in William J. Coughlin, *Conquered Press*, 116.

66 *court-martialed for publishing* and *declare any information classified*: Ibid., 122.

66 *alerted to Hersey's arrival*: "Correspondents in the Far East," Letter from F. G. Tillman, Liason Officer in Charge, to FBI Director, United States Department of Justice, Doc. 6275, Tokyo, Japan, June 10, 1946, the Records of the Federal Bureau of Investigation.

66 *a restricted topic*: According to Robert Jay Lifton and Greg Mitchell in *Hiroshima in America*, 56, "the atomic bomb was virtually a forbidden

subject in Japan. Between 1945 and 1948 only four books and one collection of poetry about the bomb were published in Japan." And according to Monica Braw in *The Atomic Bomb Suppressed* (Armonk, NY: M. E. Sharpe, Inc., 1991), 92: "For Japanese journalists, information from the bombed cities and discussions about the effects of the bomb were checked, held, suspended, or deleted for four years."

67 *another official press tour*: Robert Jay Lifton and Greg Mitchell, *Hiroshima in America*, 79.

67–68 *"remind themselves as . . ."* and *"They seemed rather . . ."*: Lindesay Parrott, "Hiroshima Builds Upon Atomic Ruins," *New York Times*, February 26, 1946, 12.

68 *"No one wants . . ."*: Joseph Julian, *This Was Radio: A Personal Memoir* (New York: The Viking Press, 1975), 149.

68 *"iron curtain"* and *"nobody knows what . . ."*: Winston Churchill, "Iron Curtain" speech, March 5, 1946, Archives of the Central Intelligence Agency: https://www.cia.gov/library/readingroom/docs/1946 -03-05.pdf.

68 *"Those Japs are . . ."*: John Hersey, "Letter from Peiping," *New Yorker*, May 4, 1946.

69 *"authorized and invited . . ."* and *fourteen day maximum imposed*: "Invitation Travel Order AGPO 144-21, General Headquarters, United States Army Forces, Pacific, May 24, 1946. John Hersey Papers, Beinecke Library, Yale University.

CHAPTER FOUR: Six Survivors

71 *twenty-four-hour ordeal*: "The Tokyo Express: A Life Photographer Takes a Ride to Hiroshima on Japan's Best Train," *Life*, October 8, 1945, 27.

71 *"seething mass of . . . ," "the air was . . . ,"* and *"coaches for Japs . . ."*: Ray C. Anderson, Ph.D, M.D., *A Sojourn in the Land of the Rising Sun* (Sun City, AZ: Elan Press, 2005), 12.

71 *"lack of . . ."*: Ibid., 324.

72 *"ruins were terribly . . ."*: Kiyoshi Tanimoto, "My Diary Since the Atomic Catastrophe, up to This Day," 187.

72 *"strange, unidentifiable odor"*: Mark Gayn, *Japan Diary* (Rutland, VT: Charles E. Tuttle Company, 1981), 268.

72 *"by one instrument . . ."*: John Hersey, "After Hiroshima: An Interview with John Hersey," *Antaeus Report*, Fall 1984, 2.

72 *"I'd seen damage . . ."* and *terrified*: John Hersey, interview with Kay Bonetti, Hersey American Audio Prose Library, 1988.

72 *"a magazine of . . ."* and *as fast as he could*: Ibid.

73 *"teeming jungles of . . ."*: Russell Brines, *MacArthur's Japan* (Philadelphia: J. B. Lippincott Company, 1948), 40.

73 *unearthed 1,000 bodies*: Mark Gayn, *Japan Diary*, 267–68.

73 *"supplies from the . . ."*: Ibid., 267; also, *corn, flour, powdered milk and chocolate*: Ursula Baatz, *Hugo Makibi Enomiya-Lasalle: Mittler zwischen Buddhismus und Christentum* (Kevelaer, Germany: Topos Taschenbücher, 2017), translated from German by Nadja Leonhard-Hooper.

73 *a typhoon and floods*: Mark Gayn, *Japan Diary*, 267.

73 *"stimulated them"*: John Hersey, "Hiroshima," *New Yorker*, August 31, 1946, 50. Hersey had in his notes a post-bomb Hiroshima botanical study conducted by four Kyoto University botanists that detailed the sorts of plants that survived and on which Hersey noted the presence of feverfew and panic grass; that the study also confirmed the way in which some flora was growing back evinced an "apparent promoting effect of the radiation," and indicated that the bomb had had a "stimulating effect for germination" in some species. Hersey would co-opt this language for "Hiroshima." (Source: "On the Influence upon Plants of the Atomc Bomb in Hiroshima on August 6, 1945, Preliminary Report," undated, Kyoto University, John Hersey Papers, Beinecke Library, Yale University.)

73 *streetcars . . . blackened corpses*: Michihiko Hachiya, M.D., *Hiroshima Diary: The Journal of a Japanese Physician, August 6–September 30,*

1945, Fifty Years Later (Chapel Hill: The University of North Carolina Press, 1995), 19.

74   *"absolutely horrible"* and *"sooner die than . . ."*: Ray C. Anderson, Ph.D, M.D., *A Sojourn in the Land of the Rising Sun*, 33.

74   *reportedly still roped off*: Dr. Ray Anderson reported that in October 1946 one of his colleagues was in Hiroshima and reported the sequestered areas. Ray C. Anderson, Ph.D, M.D., *A Sojourn in the Land of the Rising Sun*, 35–36.

74   *A Planning Conference*: A Planning Conference considering the creation of an Institute of International Amity was being advised by an American military official named Lieutenant John D. Montgomery, whom Hersey also interviewed while in Hiroshima. Lieutenant Montgomery is cited in "Hiroshima" in association with this work; his name also appears on Hersey's handwritten Japan contact list. After "Hiroshima" came out, Lieutenant Montgomery wrote to Hersey about the *New Yorker* story, recalling a "dank" evening they'd spent in Hiroshima discussing the planning program, and telling Hersey that "Hiroshima" might be the most important writing to come out of the war. Letter from Lieutenant John D. Montgomery to John Hersey, September 6, 1946, John Hersey Papers, Beinecke Library, Yale University.

74   *"bomb souvenirs," "treasure area involved . . . ," "amateur scrounger,"* and *"a small fortune . . ."*: Ray C. Anderson, Ph.D, M.D., *A Sojourn in the Land of the Rising Sun*, 37.

74   *a New Year's Day "Atom Bowl"*: *New York Times* AP report published December 29, 1945, "Atom Bowl Game Listed; Nagasaki Gridiron Will Be Site of Marines' Contest Tuesday" (dateline filed December 28), and *New York Times* U.P. report published January 3, 1946, "Osmanski's Team Wins: Sets Back Bertelli's Eleven by 14–13 in Atom Bowl Game" (dateline filed January 2, 1946).

75   *"We thought it . . ."* and *"it would be . . ."*: Colonel Gerald Sanders quoted in "Nagasaki, 1946: Football Amid the Ruins," *New York Times*, John D. Lukacs, December 25, 2005, 9.

75 *Hersey found shelter*: Kiyoshi Tanimoto, "Postscript: My Diary Since the Atomic Catastrophe," and Kiyoshi Tanimoto, "My Diary Since the Atomic Catastrophe up to This Day," 173.

75 *municipality of Ujina* and *during World War II*: Professor Kazumi Mizumoto to Lesley Blume, December 8, 2018.

75 *his own "subsistence"*: "Invitation Travel Order AGPO 144-21," General Headquarters, United States Army Forces, Pacific, May 24, 1946. John Hersey Papers, Beinecke Library, Yale University.

76 *The logs . . . inspired envy*: Kiyoshi Tanimoto, "My Diary Since the Atomic Catastrophe up to This Day," 174.

76 *Father Lassalle had been*: In "Hiroshima," Hersey spells Lassalle's surname "LaSalle." But Lassalle biographer Dr. Ursula Baatz confirms that the correct spelling is Lassalle.

76 *small, one-room barracks*: Masami Nishimoto, "History of Hiroshima: 1945–1995; Hugo Lassalle, Forgotten 'Father' of Hiroshima Cathedral," *Chugoku Shimbun*, Part 14, Article 2, December 16, 1995.

76 *studied . . . in the United Kingdom.*: Ibid.

76 *standing at the window*: Father Johannes Siemes, "Atomic Bomb on Hiroshima: Eyewitness Account of F. Siemes," 3, and Masami Nishimoto, "History of Hiroshima: 1945–1995; Hugo Lassalle, Forgotten 'Father' of Hiroshima Cathedral," *Chugoku Shimbun*, Part 14, Article 2, December 16, 1995.

76 *"This is the . . ."* and *"Now I will . . ."*: Ursula Baatz, *Hugo Makibi Enomiya-Lasalle: Mittler zwischen Buddhismus und Christentum* (Kevelaer, Germany: Topos Taschenbücher, 2017), 46–47; translated from German by Nadja Leonhard-Hooper.

77 *"freed with the . . ."*: "Atomic Bomb on Hiroshima: Eyewitness Account of F. Siemes," extended written testimony, 3. This extended Father Siemes testimonial, which appeared in edited form in *Time*'s pony edition, had been circulated by the headquarters of the United States Strategic Bombing Survey (Pacific) during its investigation of the bombings' aftermath, on December 6, 1945. The cover letter

accompanying the testimony to all divisions of the Survey stated that the "document was obtained by joint investigations of a U.S. Army and Naval Technical Mission, Japan." Both the cover letter and extended testimony are among John Hersey's "Hiroshima" materials in the John Hersey Papers, Beinecke Library, Yale University.

77 *"left to their . . .":* Ibid., 3–4.

77 *Father Lassalle introduced Hersey:* Pater Franz-Anton Neyer interview with Dr. Sigi Leonhard on behalf of Lesley Blume, January 19, 2018.

78 *"I became extremely . . . ," "they told me . . . ,"* and *"bone marrow [had] . . .":* Pater Wilhelm Kleinsorge interview, *Bayersicher Rundfunk,* "Strahlen aus der Asche," 1960; translated from German by Nadja Leonhard-Hooper.

78 Stimmen der Zeit and *he blacked out:* John Hersey, "Hiroshima," *New Yorker,* August 31, 1946, 18.

78 *worried he would die:* Father Johannes Siemes, "Atomic Bomb on Hiroshima: Eyewitness Account of F. Siemes," 3.

78 *"the oncoming flames . . .":* Ibid., 4.

78 *survive the destruction . . . :* Mr. Fukai as paraphrased by Father Johannes Siemes, Ibid., 4. *forcibly carry him away:* Ibid.

78 *heard from since:* Ibid.

79 *a violent whirlwind:* Ibid.

79 *"As far as . . . ," "The banks of . . . ," "Frightfully injured forms . . . ,"* and *naked, charred cadavers . . . :* Ibid., 5.

79 *eyes had melted:* John Hersey, "Hiroshima," *New Yorker,* August 31, 1946, 33.

80 *"All these bomb . . .":* Ibid., 54.

80 *"bomb-affected people":* "Survivors of Hiroshima and Nagasaki," Atomic Heritage Foundation, Thursday, July 27, 2017: https://www .atomicheritage.org/history/survivors-hiroshima-and-nagasaki. *introduce Hersey to other blast survivors:* Father Kleinsorge later stated that "I found him the other interview partners that appear in the book; they all came from my circle of acquaintances." (Source: Franz-Anton

Neyer interview with Dr. Sigi Leonhard on behalf of Lesley Blume, January 19, 2018.) *act as a translator*: Father Kleinsorge later stated in interviews that he had done so, including in a profile in the August 6, 1952, issue of *Asahigraph* ("First Interviews with Atomic Bomb Victims").

80 *Ushita section of Hiroshima*: Letter from Kiyoshi Tanimoto to John Hersey, May 29, 1946. John Hersey Papers, Beinecke Library, Yale University.

81 *"Temporary hall and . . ."*: Kiyoshi Tanimoto, "My Diary Since the Atomic Catastrophe up to This Day," entry dated September 18, 1945.

81 *greeted Hersey and Father Kleinsorge*: According to Reverend Tanimoto's diaries, Hersey and Father Kleinsorge came to his home on Wednesday, May 29, 1946. Ibid., 172.

81 *"I [have] contracted . . ."*: Kiyoshi Tanimoto, "My Diary Since the Atomic Catastrophe up to This Day," entry dated "September 21, 1945–October 29, 1945," 113.

81 *"Father Kleinsorge of . . ."*: Kiyoshi Tanimoto, "My Diary Since the Atomic Catastrophe up to This Day," entry dated May 29, 1946, 172.

82 *"Entirely ignorant of . . ."* and *"Up to that . . ."*: Kiyoshi Tanimoto, "Postcript: My Diary Since the Atomic Catastrophe up to This Day." John Hersey Papers, Beinecke Library, Yale University.

82 *to his kindness*: Kiyoshi Tanimoto, "My Diary Since the Atomic Catastrophe up to This Day," entry of May 29, 1946, 172.

82 *ten-page missive*: The May 29, 1945, Tanimoto letter to Hersey, along with the hand-drawn map, remain among Hersey's papers in the Beinecke Library, Yale University.

82 *"unlike a soldier . . ."*: Kiyoshi Tanimoto, "Postcript: My Diary Since the Atomic Catastrophe."

82–83 *"I've read your . . . ," "from the standpoint . . . ,"* and *"listened to me . . ."*: Kiyoshi Tanimoto, "My Diary Since the Atomic Catastrophe up to This Day," entry of May 29, 1946, 173.

83   *5:00 a.m. on August 6, 1945*: Kiyoshi Tanimoto, "My Diary Since the Atomic Catastrophe up to This Day," entry of May 29, 1946, 1–2.

83   *"fantastic rumors that . . ."*: Father Johannes Siemes, "Atomic Bomb on Hiroshima: Eyewitness Account of F. Siemes," 1.

83   *fire prevention lanes*: Kiyoshi Tanimoto, "My Diary Since the Atomic Catastrophe up to This Day," entry of May 29, 1946, 3.

83   *Bibles, church records, and altar objects*: Ibid., 1.

83   *"sharp flash of . . ."*: Ibid., 4.

83   *fell to the ground*: Ibid., 4–5.

84   *"Most of the . . ."* and *"procession of ghosts"*: Ibid., 8–9.

84   *"were all flat . . ."*: Ibid., 10.

84   *"painful cries of . . ."*: Ibid.

84   *"Red hot iron . . ."*: Ibid., 20.

85   *"[Chisa] was in . . ."*: Ibid., 15.

85   *"she struggled with . . ."*: Ibid., 17.

85   *scratch out a hole* and *emerged just in time*: Koko Tanimoto Kondo interview with Lesley Blume, November 29, 2018.

85   *clinging to her dead infant* and *it began to decompose*: Kiyoshi Tanimoto, "My Diary Since the Atomic Catastrophe up to This Day," 22, 42.

85–86   *saw a small boat* and *"I had no . . ."* and details of ferrying survivors to Asano Park: Ibid., 22–23.

86   *"the skin of . . ."*: Ibid., 34.

86   *"Mother, I'm cold . . ."*: Ibid., 36.

86   *Dr. Fumio Shigeto* and *already been exposed*: Ibid., 1.

86   *walked . . . into the neighborhood*: Ibid., 173.

87   *"an acquaintance I . . ."*: Ibid., 173–74.

87   *taking notes throughout*: Hersey later told interviewer David Sanders that he had taken notes during these interviews, which were unrecorded. ("He interviewed his subjects before tape recorders existed, before any kind of recording device was standard equipment for

investigative reporters, and therefore expanded and arranged his hand-written notes . . .": Hersey interview with Sanders, August 13, 1987, as summarized in *John Hersey Revisited*, David Sanders [Boston: Twayne Publishers, 1991], 15.) Hersey had previously used small notebooks while on reporting assignments; in Guadalcanal, he had sheathed his notebooks in condoms to keep them dry. (Source: John Hersey, *Into the Valley*, xx.) However, his Hiroshima notes from his May and June 1946 interviews are not among his papers at the Beinecke Library at Yale University, which otherwise contains other research materials he consulted when writing the article.

87   *may have used shorthand*: Hersey later recalled that "Lewis gave me a month to learn shorthand—he suggested either the Gregg system or speed-writing—and to switch over from hunt-and-peck to touch typing." (Source: John Hersey, "First Job," *Yale Review*, Spring 1987, as reprinted in John Hersey, *Life Sketches* [New York: Alfred A. Knopf, 1989], 13.) With the Gregg system, one could reportedly record by hand up to 225 words a minute. For more information, see Dennis Hollier, "How to Write 225 Words per Minute with a Pen," *Atlantic*, June 24, 2014: https://www.theatlantic.com/technology/archive/2014/06/yeah-i-still-use-shorthand-and-a-smartpen/373281/.

88   *dwelling in a one-room shack* and *Meager meals*: Norman Cousins, "John Hersey," *Book of the Month Club News*, March 1950.

88   *"He sat on . . ."*: Mrs. Hatsuyo Nakamura, as quoted in Norman Cousins, "John Hersey," *Book-of-the-Month Club News*, March 1950.

88   *The moment that the bomb exploded*: John Hersey, "Hiroshima," *New Yorker*, August 31, 1946, 17.

89   *naval hospital ship*: Kiyoshi Tanimoto, "My Diary Since the Atomic Catastrophe up to This Day," 31.

89   *270 of the city's 300 doctors*: The cited Hiroshima doctor and nurse casualty count as per Dr. Marcel Junod of the International Committee of the Red Cross, who surveyed Hiroshima in September 1945 and recorded these casualty statistics in his diary. Diary contents

summarized and excerpted on ICR website: "The Hiroshima Disaster—a Doctor's Account," December 9, 2005: https://www.icrc.org/en/doc/resources/documents/misc/hiroshima-junod-120905.htm. In "Hiroshima," Hersey would state that 65 of 150 Hiroshima doctors had died, and that most of the rest were wounded. His statistic for the nurse casualty rate is the same. (Source: John Hersey, "Hiroshima," *New Yorker*, August 31, 1946, 21.)

89 *150 glass shards*: Michihiko Hachiya, M.D., *Hiroshima Diary: The Journal of a Japanese Physician, August 5–September 30, 1945, Fifty Years Later* (Chapel Hill: The University of North Carolina Press, 1995), v.

89 *poison gas or perhaps some deadly micoorganisms*: Ibid., 21.

90 *three Japanese interpreters*: It is unclear who the additional three Japanese translators were, but another source indicates that there were quite a few English speakers around that time in Hiroshima: "There are a great number of Japs in Hiroshima who have spent time in the States," said one American doctor, who was based in Hiroshima not long after Hersey's trip and encountered during his work in the city many American-educated and American-born nisei, or people born in the U.S. to Japan-born parents. (Source: Ray C. Anderson, Ph.D, M.D., *A Sojourn in the Land of the Rising Sun*, 229.)

90 *"came as an . . ."* and *"packed, like rice . . ."*: Michihiko Hachiya, M.D., *Hiroshima Diary: The Journal of a Japanese Physician, August 5–September 30, 1945, Fifty Years Later* (Chapel Hill: The University of North Carolina Press, 1995), 11–12.

91 *"M. Fujii, M.D. . . ."*: John Hersey, "Hiroshima," *New Yorker*, August 31, 1946, 58.

91–92 *"Some were doctors . . ."* and *"When he was . . ."*: Dr. Masakazu Fujii, as quoted in Norman Cousins, "John Hersey," *Book-of-the-Month Club News*, March 1950.

92 *"like a morsel . . ."*: John Hersey, "Hiroshima," *New Yorker*, August 31, 1946, 18.

93 *Reverend Tanimoto grew angry* and *get an army doctor*: Kiyoshi Tanimoto, "My Diary Since the Atomic Catastrophe up to This Day," 38–39.

93 *took counsel and received solace*: John Hersey, "Hiroshima," *New Yorker*, August 31, 1946, and "First Interviews with Atomic Bomb Victims," *Asahigraph*, August 6, 1952, translated from Japanese by Ariel Acosta.

94 *"three grotesques"*: John Hersey, "Hiroshima," *New Yorker*, August 31, 1946, 24.

95 *"by no means . . ."*: John Hersey, "John Hersey: The Art of Fiction No. 92," interview by Jonathan Dee, *Paris Review*, issue 100, Summer–Fall 1986.

95 *military police officers . . . knew*: Kiyoshi Tanimoto, "My Diary Since the Atomic Catastrophe up to This Day," 173.

96 *"hanged by the . . ."*: Mark Gayn, *Japan Diary*, 258.

96 *stood on the platform* and *have to write a line*: John Hersey, American Audio Prose Library, interview with Kay Bonetti, 1988.

96 *memorial services*: Kiyoshi Tanimoto, "My Diary Since the Atomic Catastrophe up to This Day," 179.

96 *entertained occupationaires*: John Hersey, "Hiroshima," *New Yorker*, August 31, 1946, 58.

96 *hospital in Tokyo*: Ibid., 64, and Pater Franz-Anton Neyer interview with Dr. Sigi Leonhard on behalf of Lesley Blume, January 19, 2018.

96 *hair was growing back* and *back in school*: John Hersey, "Hiroshima," *New Yorker*, August 31, 1946, 63, and Norman Cousins, "John Hersey," *Book-of-the-Month Club News*, March 1950.

CHAPTER FIVE: Some Events at Hiroshima

97 *Their latest issue*: *New Yorker*, June 15, 1946.

97 *cable from Hersey*: John Hersey cable to William Shawn, received June 12, 1946, 3:30 p.m., *New Yorker* records, New York Public Library.

97 *interminable Air Training Command flight*: Letter from Elizabeth Gilmore to John Hersey, September 7, 1946, John Hersey Papers, Box

36, Beinecke Library, Yale University, and Hersey notes, John Hersey Papers, Beinecke Library, Yale University.

98 *Hickam Field* and *Thirty-five men died instantly*: "Hickam Field," *Aviation: From Sand Dunes to Sonic Booms*, U.S. Department of the Interior, National Park Service website: https://www.nps.gov/articles /hickam-field.htm (as of March 14, 2019).

98 *"white heat"*: John Hersey, "John Hersey: The Art of Fiction No. 92," interview by Jonathan Dee, *Paris Review*, issue 100, Summer–Fall 1986.

98 *possible title for the story*: John Hersey, Manuscript, First Draft, "Hiroshima," John Hersey Papers, Beinecke Library, Yale University.

98 *" 'Original Child' Bomb"*: John Hersey, "Hiroshima," *New Yorker*, August 31, 1946, 43.

98 *"Journalism allows its . . ."*: John Hersey, "The Novel of Contemporary History," *Atlantic Monthly*, 1949.

98 *"to have the . . ."*: John Hersey interview with Kay Bonetti, American Audio Prose Library, 1988.

99 *"My choice was . . ."*: John Hersey, "John Hersey: The Art of Fiction No. 92."

99 *"suppression of horror"* and *"an effect far . . ."*: John Hersey to Michael Yavenditti, July 30, 1971, as quoted in "Hersey and the American Conscience: The Reception of 'Hiroshima,' " *Pacific Historical Review* Vol. 43, No. 1 (Feb., 1974).

99 *(military-approved) book*: In June, 1946, William Laurence was preparing to release *Dawn Over Zero*; the book was released on August 22 of that year. (Source: "Book Notes," *New York Herald Tribune*, July 23, 1946.)

99 *"the hills said . . ."*: William L. Laurence, *Dawn Over Zero* (New York: Alfred A. Knopf, 1946), 4.

100 *"At exactly fifteen . . ."*: John Hersey, "Hiroshima," *New Yorker*, August 31, 1946, 15.

100 *"And now each . . ."*: Ibid.

101 *"There in the . . ."*: John Hersey, Manuscript, First Draft, "Hiroshima," John Hersey Papers, Beinecke Library, Yale University.

101 *[Tanimoto's] family*: In the draft of "Hiroshima," the published article, and the book version, Hersey misidentified Reverend Tanimoto's infant, Koko, as a son.

101 *"an automaton, mechanically . . ."*: John Hersey, "Hiroshima," *New Yorker*, August 31, 1946, 22.

102 *cache of post-bomb Japanese studies*: Hersey also had among his reference materials a 1926 report by a Japanese doctor, Dr. Masao Tsuzuki, titled "Experimental Studies on the Biological Action of Hard Roentgen Rays," which listed the results of experiments studying the effects of radiation on laboratory animals. Hersey may also have heard about the report or been given the report by *New Yorker* writer Daniel Lang, who had just written about Dr. Tsuzuki in a story for the magazine titled "A Fine Moral Point" (June 8, 1946). In the article, Lang quoted Dr. Tsuzuki as saying, "It remained for [the Americans] to conduct the human [radiation] experiment" (page 64).

102 *"General MacArthur's headquarters . . ."*: John Hersey, Manuscript, First Draft, "Hiroshima," John Hersey Papers, Beinecke Library, Yale University.

103 *"[But] trying to keep . . ."*: Ibid.

104 *"as if nature . . ."*: John Hersey, "Hiroshima," *New Yorker*, August 31, 1946, 58.

104 *a study conducted by*: Hiroshima City's casualty statistics report was among Hersey's "Hiroshima" writing materials. The report stated that as of November 30, 1945, 78,150 Hiroshima civilians had died and 13,983 were missing. An additional 9,428 had been seriously wounded and 27,997 had been "slightly wounded." (Source: "Statistics of Damages Caused by Atomic Bombardment, August 6, 1945," Foreign Affairs Section, Hiroshima City, undated. John Hersey Papers, Beinecke Library, Yale University.)

104 *However, no one* and *270,000 dead and wounded*: "A Preliminary

Report on the Disaster in Hiroshima City Caused by the Atomic Bomb," Research Commission of the Imperial University of Kyoto, John Hersey Papers, Beinecke Library, Yale University.

105 *"the exact number . . ."*: "U.S. Strategic Bombing Survey: The Effects of the Atomic Bombings of Hiroshima and Nagasaki," June 19, 1946: https://www.trumanlibrary.org/whistlestop/study_collections/bomb /large/documents/pdfs/65.pdf, and John Hersey Papers, Beinecke Library, Yale University.

105 *Truman had requested the study*: Ibid.

105 *"fairly full account . . ."*: Ibid.

105–6 *"problems of defense," "What if the . . . ," "The danger is . . . ," "value of decentralization," "crippled or wiped . . . ," "wise zoning," "reshaping and partial . . . ,"* and *"our understanding of . . ."*: Ibid.

107 *"But Kikuki's mother . . ."*: John Hersey, "Hiroshima," *New Yorker*, August 31, 1946, 68.

107 *"Look, we just . . ."*: William Shawn as quoted by Hersey in "John Hersey: The Art of Fiction No. 92."

107 *"This can't be . . ."*: "The Press: Six Who Survived," *Newsweek*, September 9, 1946, 70.

107 *"an unprecedented editorial . . ."*: John McPhee email to Lesley Blume, January 26, 2018.

108 *"cheated"*: Harold Ross biographer Thomas Kunkel email to Lesley Blume, November 15, 2018, and "The Press: Six Who Survived," *Newsweek*, September 9, 1946, 70.

108 *resolved any qualms*: On the day of the Pearl Harbor attack, a Sunday, Ross and Shawn both beelined for the *New Yorker* offices and immediately "[put] the magazine on what Ross called a war basis," recalled *New Yorker* contributor James Thurber, "making over 'Talk of the Town,' ripping out civilian ornaments and replacing them with spots of cannon and of flags, sending reporters scurrying all over for war features, profile ideas, and Reporter pieces. (Source: James Thurber, *The Years with Ross* [Boston: Little, Brown and Company, 1959], 166.)

108   *"having gone heavyweight . . ."*: Harold Ross letter to Janet Flanner, June 25, 1946, as quoted in Ben Yagoda, *About Town* (New York: Scribner, 2000), 193.

108   *"Hersey has written . . ."* and *"He wants to . . ."*: Harold Ross letter to E. B. White, August 7, 1946, as quoted in Ben Yagoda, *About Town*, 183.

109   *"gay, humorous, [and] . . . ," "with a declaration . . . ," "publish facts that . . . ,"* and *"try conscientiously to . . ."*: "Of All Things," *New Yorker*, February 21, 1925, 2.

109   would *"present the . . ."*: Harold Ross, "The *New Yorker* Prospectus," Fall 1924, as reprinted in Thomas Kunkel, *Genius in Disguise: Harold Ross of The New Yorker* (New York: Carroll & Graf Publishers, Inc., 1996), 439–41.

109   *exactly what Ross needed*: "The Press: Six Who Survived," *Newsweek*, September 9, 1946, 70.

109   *called Shawn and Hersey*: Thomas Kunkel, *Genius in Disguise*, 372, and John Hersey, "John Hersey: The Art of Fiction No. 92."

109   *"After a couple . . ."* and *"I don't know . . ."* : Harold Ross letter to Rebecca West, August 27, 1946, as quoted in Thomas Vinciguerra, *Cast of Characters: Wolcott Gibbs, E. B. White, James Thurber, and the Golden Age of The New Yorker* (New York: W. W. Norton & Company, 2016), 285.

109   *a copy of* Who's Who: Brendan Gill, *Here at* The New Yorker (New York: Random House, 1975), 42.

110   *10:00 a.m. . . . until 2:00 a.m.*: "The Press: Six Who Survived," *Newsweek*, September 9, 1946, 70.

110   *"gobsmacked"*: Thomas Kunkel interview with Lesley Blume, November 14, 2018.

110   *a "dummy" issue*: No copy of a "dummy" issue—or "alternative" August 31, 1946 issue—appears to exist among the *New Yorker* records at the New York Public Library, but several retired *New Yorker* employees recall being told that a fake issue had been in play, and that besides

Ross, Shawn, Hersey, and Ross's secretary, only "makeup" (or layout) man Carmen Peppe knew about the "Hiroshima" issue being edited in Ross's office. One former *New Yorker* makeup employee recalled being told by previous members of the department that "only a handful of editorial staff members knew about the [Hersey] article, and that another alternative version of that particular issue was being approved to maintain secrecy." (Source: Pat Keogh email to Lesley Blume, February 9, 2018). Another longtime *New Yorker* editor recalled: "I heard [the] story from the guys in the makeup department, who'd been there since the 1920s . . . There was a fake issue of the magazine that everyone worked on. Only Shawn and Ross and the head of the makeup department knew about it. This sounded preposterous to me, but they swore this was the case." (Source: John Bennet interview with Lesley Blume, February 7, 2018.)

110 *"There would have . . ."* : Thomas Kunkel interview with Lesley Blume, November 14, 2018.

110 *secretly stashed aside* and *the* New Yorker's *business office*: "The Press: Six Who Survived," *Newsweek*, September 9, 1946, 70. *Chesterfield cigarettes, Perma-Lift brassieres . . .* : See *New Yorker*, August 31, 1946.

111 *"Ross could never . . ."*: Brendan Gill, *Here at* The New Yorker, 181.

111 *"a strong aversion . . ."* and *"as a measure . . ."*: Ibid., 25.

111 *"a very fine . . . ," "practically everything," the definitive piece . . . , "How many were . . . ," "trouble with these . . . ,"* and *"the hour or minute . . ."*: Harold Ross memo on Part II of "Some Events at Hiroshima," August 6, 1946, as reprinted in Ben Yagoda, *About Town*, 190.

112 *"I don't see . . . ," "an old air . . . ," "slender," "could one row," "he said this . . . ,"* and *"I'll be damned . . ."*: Ibid., 188–89.

112 *tormented in the early hours*: James Thurber, *The Years with Ross*, 167.

112 *settled on "crumpled"* and *same word down*: John Hersey, "John Hersey: The Art of Fiction No. 92."

112 *"a kind of . . ."* and *"to think in . . ."*: Ibid.

113 *"If you're from . . ."*: Adam Gopnik interview with Lesley Blume, June 15, 2017.

113 *"Japan Notes Atom . . ."* and *"few signs have . . ."* : Lindesay Parrott, "Japan Notes Atom Anniversary; Hiroshima Holds Civic Festival," *New York Times*, August 7, 1946, 13.

113 *"thousands of . . . [Hiroshima] . . . ,"* *"bangup ritual lantern . . . ,"* and *"stampeded in the . . ."*: "Japan: A Time to Dance," *Time*, August 19, 1946, 36.

113 *"There, in the . . ."*: John Hersey, "Hiroshima," *New Yorker*, August 31, 1946, 19.

114 *"the most sensational . . ."*: Harold Ross letter to John Hersey, September 11, 1946, *New Yorker* records, New York Public Library.

114 *"going to bear . . ."*: Harold Ross letter to John Hersey, September 25, 1946, *New Yorker* records, New York Public Library.

115 *submitted their war stories*: The *New Yorker* records at the New York Public Library contain many wartime exchanges between *New Yorker* editors and War Department public relations officers pertaining to censorship/clearance submissions.

115 *terminating the wartime Office of Censorship*: Executive order 9631 ("TERMINATION OF THE OFFICE OF CENSORSHIP(1)," signed on September 28, 1945, decreed that "The Office of Censorship, established by Executive Order No. 8985 of December 19, 1941, shall continue to function for the purposes of liquidation until the close of business on November 15, 1945, at which time the Office (including the office of the Director of Censorship) shall terminate." (Source: Executive Orders, Harry S. Truman, 1945–1953, Harry S. Truman Presidential Library & Museum website, https://www.trumanlibrary.org /executiveorders/index.php?pid=391&st=&st1=.)

115 *the government's confidential order*: September 14, 1945, government letter to editors: Monica Braw, *The Atomic Bomb Suppressed* (Armonk, NY: M. E. Sharpe, Inc., 1991), 111.

115 *"spot check . . ."*: Daniel Lang, "A Fine Moral Point," *New Yorker*, June 8, 1946, 62.

115 *"in the matter . . ."* and *"this and past . . ."* : Sanderson Vanderbilt letter to Major Walter King, May 15, 1946, *New Yorker* records, New York Public Library. Sanderson had been a *New Yorker* employee before the war, but during the war had served with the U.S. Army, writing for its publication, *Yank*; he would have been an excellent postwar conduit to the War Department's public relations operation for the *New Yorker*, whose staff he rejoined in 1945. (Source: "Sanderson Vanderbilt, 57, Dies; *New Yorker* Editor Since 1938," *New York Times*, January 24, 1967, 28.)

115–16 *"restricted data," "all data concerning . . . ," "with any reason . . . ," "to injure the . . . ," "secure an advantage . . . ,"* and *"be punished by . . ."* : Atomic Energy Act of 1946, Public Law 585, 79th Congress, Chapter 724, 2D Session, S. 1717.

116 *"Should we submit . . . ," "Mr. Shawn and . . . ," "has come from . . . ," "the Army supplied . . . ,"* and *"find out where . . ."*: Harold Ross letter to Milton Greenstein, August 1, 1946, *New Yorker* records, New York Public Library.

116–17 *" 'Data' is not . . . ," "of course, we . . . ," "there may be . . . ,"* and *"probably ought not . . ."*: Milton Greenstein letter to Harold Ross, August 12, 1946, *New Yorker* records, New York Public Library.

117 *"even be enough . . ."*: Michael Sweeney email to Lesley Blume, March 24, 2019.

118 *"four-part article on . . ."*: William Shawn letter to General Leslie Groves, August 15, 1946, *New Yorker* records, New York Public Library.

118 *"As I look . . ."*: General Leslie Groves speech to IBM Luncheon, September 21, 1945, as quoted in "Keep Bomb Secret, Gen. Groves Urges," *New York Times*, September 22, 1945, 3. Harold Ross and William Shawn likely would have been aware of General Groves's remarks and attitude. This particular statement, for example—delivered

to a room of two hundred people the previous autumn at a luncheon held in his honor at the Waldorf-Astoria hotel—had been reported prominently in the *New York Times*, as were similar General Groves speeches and remarks at other events that autumn.

119 *"If there was . . . ,"* and *"we must have . . ."*: General Leslie Groves memo to John M. Hancock, January 2, 1946, as excerpted in Robert S. Norris, *Racing for the Bomb* (South Royalton, VT: Steerforth Press L.C., 2002), 472–73.

119 *start to see an urgent need*: Mainstream publications were already helping Americans envision their country thus threatened; for example, *Newsweek* would shortly run a story about plans being circulated for vast underground shelters and even underground villages that could be built underneath cities. (Source: "Ever-Ever Land?" *Newsweek*, September 9, 1946, 66.)

120 *"changing the article . . ."* and *"would not hurt . . ."*: Notes from General Leslie R. Groves's appointment book notes, entry August 7, 1946, National Archives and Records Administration, and courtesy of personal files of Groves biographer Robert S. Norris. In the conversation, Groves proposed dispatching either a "Col. Derry" or a "Maj. Coakley" to the *New Yorker* offices. William Shawn requested that the officer come the following morning, "as if there were any vital changes they could make them before afternoon." It is unclear which officer ultimately went, although a Major Robert J. Coakley did have a subsequent correspondence with Shawn after this date. Ibid.

120 *details of the . . . meeting are unknown*: Hersey may have been in attendance in the meetings with General Groves's representative, although later he did not disclose in later interviews or in his retellings of the making of "Hiroshima" that such any such meeting had taken place. However, there is evidence that he did, at the very least, know that "Hiroshima" had been subjected to review by the War Department. Five months later, when corresponding with the War Department to lobby for publication of "Hiroshima" in Japan, he noted that the story had

been cleared by the War Department. (Source: John Hersey letter to Jay Cassino, January 8, 1947, John Hersey Papers, Beinecke Library, Yale University.)

120 *"ten times as . . ."*: John Hersey, "Hiroshima," *New Yorker*, August 31, 1946, 62.

121 *"to look over"*: William Shawn letter to General Leslie R. Groves, August 15, 1946, *New Yorker* records, New York Public Library.

122 *worked for the Office of War Information*: "Charles Elmer Martin, or CEM, New Yorker Artist, Dies at 85," *New York Times*, June 20, 1995, B10.

122 *"escape[d] into easy . . ."*: Albert Einstein, speech at Hotel Astor, Nobel fifth anniversary, "The War Is Won, but the Peace Is Not," December 10, 1945, as quoted in David E. Rowe and Robert Schulmann, eds., *Einstein on Politics: His Private Thoughts and Public Stands on Nationalism, Zionism, War, Peace, and the Bomb* (Princeton, NJ: Princeton University Press, 2007), 381–82.

122 *"My God, how . . ."*: An unnamed *New Yorker* editor quoted in a story about the making of "Hiroshima" in *Time*: "Without Laughter," *Time*, September 9, 1946.

123 *white paper bands*: Ibid.

123 *a distracting failure*: John Hersey letter to Alfred Knopf, September 6, 1946, Alfred A. Knopf, Inc., Archive, Harry Ransom Center, University of Texas at Austin.

123 *Photographs . . . made available . . . via Acme*: R. Hawley Truax letter to Edgar F. Shilts, September 13, 1946, *New Yorker* records, New York Public Library. The *New Yorker* would not include photography to accompany its features until the 1990s, under editor Tina Brown. (Source: Natalie Raabe, executive director and head of communications, *New Yorker*, email to Lesley Blume, April 8, 2019.)

123 *"Like other Americans . . ."*: John Hersey interview with Kay Bonetti, American Audio Prose Library interview, 1988.

123 *reprint proceeds to the American Red Cross*: "Memorandum on the

Use of the Hersey Article," August 30, 1946, *New Yorker* records, New York Public Library. Regarding all parties agree to the announcement: "All in abeyance" noted on memorandum.

123 *personally ferried the article*: "The Press: Six Who Survived," *Newsweek*, September 9, 1946, 70, and John Bennet interview with Lesley Blume, February 7, 2018.

124 *"I feel as . . ."*: Harold Ross note, undated, John Hersey papers, Beinecke Library, Yale University.

124 *"Dear John, . . ."*: William Shawn letter to John Hersey, August 27, 1946, John Hersey Papers, Beinecke Library, Yale University.

CHAPTER SIX: Detonation

125 *Thursday, August 29, 1946*: Letter from *New Yorker* editors to news outlet editors regarding "Hiroshima," August 28, 1946, *New Yorker* records, New York Public Library.

125–26 *"I hurried over . . ."* and *"I thought that . . ."*: Lillian Ross, *Here but Not Here* (New York: Random House, Inc., 1998), 155. *small, spare office*: Ibid., 21–22.

125 *"explosive . . ."*: John Hersey, "John Hersey: The Art of Fiction No. 92," *Paris Review*, interview by Jonathan Dee, issue 100, Summer–Fall 1986.

126 *"TO OUR READERS . . ."*: Editor's note introducing "Hiroshima," *New Yorker*, August 31, 1946, 15.

126 *"broken a precedent . . ."* and *"a terrifically important . . ."*: Letter from *New Yorker* editors to news outlet editors regarding "Hiroshima," August 28, 1946, *New Yorker* records, New York Public Library.

127 *"gone out confidently"* and *"gone out on . . ."*: Harold Ross letter to Charles Merz, September 5, 1946, *New Yorker* records, New York Public Library.

127 *Hersey left New York City*: Harold Ross was left to explain to the press and others that "[Hersey] left town after finishing up his proofs on

the piece and hasn't been back here since, and isn't coming soon."
(Source: Letter from Harold Ross to Jack Skirball, September 12,
1946, *New Yorker* records, New York Public Library.) Hersey's correspondence during the period of the article's release up until September
27 bears the Blowing Rock, North Carolina, address.

127 *"ducked out of . . . ,"*: "The Press: Atomic Splash," *Newsweek*, September 9, 1946, 70. *"The response to . . ."*: *Watertown Daily Times*,
September 9, 1946, 7.

127 *"We managed to . . ."*: Richard Pinkham letter to Harold Ross, September 11, 1946, *New Yorker* records, New York Public Library.

128 *the best reporting . . . ,* and *"You smell the . . ."*: Lewis Gannett, "Books
and Things," *New York Herald Tribune*, August 29, 1946, 23.

128 *"old paradox which . . . ,"* and *"the tragedy of . . ."*: Editorial, *New York
Herald Tribune*, August 30, 1946, 14.

128 *"We are hearing . . ."*: Harold Ross letter to Lewis Gannett, September
11, 1946, *New Yorker* records, New York Public Library.

128 *most widely reprinted work*: Harold Ross letter to Jack Skirball, September 12, 1946, *New Yorker* records, New York Public Library.

129 *"It is what . . ."* and *"HIROSHIMA—DEATH OF . . ."*: "Editorial for
Peace," *Indianapolis News*, September 10, 1946.

129 *"This article is . . . ,"* *"to conceal from . . . ,"* *"amoral fools,"* and
"tolerate no [more] . . .": "News and Comments," *Monterey Peninsula
Herald*, September 10, 1946.

129 *"swell job"*: Secretarial note to Harold Ross, August 29, 1946, 1:00
p.m., *New Yorker* records, New York Public Library.

129–30 *small item* and *cartoons were conspicuously absent*: "Atom Bomb
Edition Out," *New York Times*, August 29, 1946.

130 *"Every American who . . . ,"* *"The disasters at . . . ,"* *"the death
and . . . ,"* and *"history is history . . ."*: "Time from Laughter," *New
York Times*, August 30, 1946, 13.

131 *"I bow, deeply"*: J. Markel letter to Harold Ross, August 30, 1946,
John Hersey Papers, Beinecke Library, Yale University.

131 *if an article like "Hiroshima"* . . . : Don Hollenbeck letter to William Shawn, September 9, 1946, *New Yorker* records, New York Public Library.

131 *one from London's* Daily Express: C. V. R. Thompson letter to Harold Ross, August 30, 1946, *New Yorker* records, New York Public Library.

132 *"Quite a stunt . . ."*: Letter from Dick [no surname] to John Hersey, undated, John Hersey Papers, Beinecke Library, Yale University.

132 *headier than hell*: Harold Ross letter to Charles Merz, September 5, 1946, *New Yorker* records, New York Public Library.

132 *"The story is . . ."*: Harold Ross letter to Kay Boyle, September 5, 1946, *New Yorker* records, New York Public Library.

132 *"a bigger success . . ."*: Harold Ross letter to Janet Flanner, November 25, 1946, *New Yorker* records, New York Public Library.

132 *hadn't been so satisfied*: Harold Ross letter to Blanche Knopf, September 5, 1945, *New Yorker* records, New York Public Library.

132 *"No More* New Yorkers*"*: Louis Forster memo to Harold Ross, William Shawn, and John Hersey, September 5, 1946, *New Yorker* records, New York Public Library.

132 *"people rush up . . ."* and *"could probably get . . ."*: William McGuire memo to William Shawn, August 30, 1945, *New Yorker* records, New York Public Library.

132 *offered to him for $6*: Gordon Weel letter to John Hersey, August 1946, John Hersey Papers, Beinecke Library, Yale University.

132 *Japanese American soldiers buying copies*: William McGuire memo to William Shawn, August 30, 1945, *New Yorker* records, New York Public Library.

133 *deluge of Hiroshima . . .*: Louis Forster Jr. memo to Harold Ross, December 13, 1946, *New Yorker* records, New York Public Library. A UCLA study of reader letters found that "the geographic origin of the writers was diverse," hailing from "a wide range of rural and urban

areas," including many major cities, the South, and the Northwest region. (Source: Joseph Luft, "Reaction to John Hersey's 'Hiroshima' Story," UCLA report, July 14, 1947, *New Yorker* records, New York Public Library.)

133   *tasked with keeping track*: Louis Forster to Harold Ross, September 6, 1946, September 9, 1946, and September 17, 1946, *New Yorker* records, New York Public Library.

133   *George R. Caron . . . called the* New Yorker: Louis Forster memo to Harold Ross, William Shawn, and John Hersey, September 4, 1946, *New Yorker* records, New York Public Library.

133   All letter excerpts to Ross, Shawn, and Hersey are as quoted in Joseph Luft, "Reaction to John Hersey's 'Hiroshima' Story," UCLA report, July 14, 1947, *New Yorker* records, New York Public Library. This UCLA study of the incoming "Hiroshima"-related mail sent to Hersey and the *New Yorker* concluded that "the impact of this story on the American people was striking" and that the overwhelming majority of surveyed letters "expressed unqualified approval of the article."

134   *"Wonderful—marvelous. Now . . .":* J. B. Betherton letter to John Hersey, September 1946, Beinecke Library, Yale University.

134   *stunt and, "propaganda aimed at . . . ,"* and *"Had we lost . . .":* Editorial, *New York Daily News*, September 16, 1946.

135   *stopped reading it halfway through*: Dwight MacDonald, "Hersey's Hiroshima," *politics*, October 1946, 308.

135   *"the marvelous and . . .":* Mary McCarthy to Dwight MacDonald, *politics*, November 1946, 367.

135   *"Notoriously the editors . . .":* Book-of-the-Month pamphlet, Fall 1946, Alfred A. Knopf, Inc., Archive, Harry Random Center, University of Texas at Austin.

135   *three-page behind-the-scenes "Hiroshima" story*: "The Press," *Newsweek*, September 9, 1946, 69–71.

135   *"two of them . . . ,"* and *"the bastards were . . . ,":* Harold Ross letter

to Hersey, September 11, 1946, *New Yorker* records, New York Public Library.

136 *"At 21 years . . . ,"* *"doomsday documentary,"* and *"Editor Ross, admitting . . ."*: "Without Laughter," *Time*, September 9, 1946.

136 *ingrate prodigal son*: Theodore H. White, *In Search of History* (New York: Harper & Row, Publishers, Inc., 1978), 258.

136 *Hersey's portrait removed*: Thomas Kunkel, *Genius in Disguise* (New York: Carroll & Graf Publishers, Inc., 1996), 374.

136 *"done straight"*: Karen Fishman, David Jackson, and Matt Barton, "John Hersey's 'Hiroshima' on the Air: The Story of the 1946 Radio Production," Library of Congress, October 6, 2016: https://blogs.loc .gov/now-see-hear/2016/10/john-herseys-hiroshima-on-the-air-the -story-of-the-1946-radio-production/.

137 *"I welcomed any . . ."* and *"I knew the . . ."*: Joseph Julian, *This Was Radio: A Personal Memoir* (New York: The Viking Press, 1975), 155.

137 *"This chronicle of . . ."*: "Hiroshima Report," American Broadcasting Company, Inc., September 9, 1946, 9:30–10:00 p.m.

137 *switchboards were swamped*: Robert Saudek letter to John Hersey, September 13, 1946, John Hersey papers, Beinecke Library, Yale University.

137 *highest rating of any public interest broadcast*: Robert Saudek letter to R. Hawley Truax, October 25, 1946, *New Yorker* records, New York Public Library.

137 *"scoop of the year"*: "Award Profile: *Hiroshima*, 1946, ABC Radio, Robert Saudek (Honorable Mention), Outstanding Education Program," Peabody website: http://www.peabodyawards.com/award-pro file/hiroshima.

137 *(BBC) also aired the adaptation*: The British Broadcasting Corporation aired the "Hiroshima" adaptation on four consecutive nights: October 14, 15, 16, and 17, 1946. (Source: F. S. Norman letter to Harold Ross, October 7, 1946, John Hersey Papers, Beinecke Library, Yale University.) *five hundred other U.S. radio stations*: "A Survey of Radio

Comment on the Hiroshima Issue of THE NEW YORKER," Radio Reports Inc., September 6, 1946. The survey monitored radio coverage from August 28 through September 5, and stated that "at least half of the 1,000 U.S. stations carried the story."

138 *"As I read . . . ," "all too easy . . . ,"* and *"and then maybe . . ."*: Bill Leonard, "This Is New York," WABC, August 30, 1946.

138 *"not a few . . ."*: Raymond Swing, WJZ ABC, August 30, 1946.

138 *"I will never . . ."* and *"I will never . . ."*: Ed and Pegeen Fitzgerald, "The Fitzgeralds," WJZ NYC, August 31, 1946.

138 *"Mr. Brains" and "Mrs. Beauty"*: Alec Cumming and Peter Kanze, *New York City Radio* (Charleston, SC: Arcadia Publishing, 2013), 59.

138 *continued to trouble him*: Charles J. Kelly, *Tex McCrary: Wars, Women, Politics; An Adventurous Life Across the American Century*, 109.

139 *"You know, Tex . . ."* and *"In one sense . . ."*: Jinx Falkenburg and Tex McCrary, "City of Decision," WEAF, September 4, 1946, 6:15 p.m.

139 *"more than a . . ."*: Tex McCrary, *Hi Jinx*, WEAF, August 30, 1946, 8:30 a.m.

140 *"I covered it . . ."*: Tex McCrary, as quoted in Richard Severo, "Tex McCrary Dies at 92; Public Relations Man Who Helped Create Talk-Show Format," *New York Times*, August 30, 2003.

140 *"had come swarming. . ."*: Editorial, *New York Daily News*, September 16, 1946.

140 *"for the time . . ."*: Memo from R. Hawley Truax to John Hersey, Harold Ross, and William Shawn, August 29, 1946, *New Yorker* records, New York Public Library.

140 *several offers*: Harold Ross letter to Jack Skirball, September 12, 1946, *New Yorker* records, New York Public Library.

140 *"Ten Outstanding Celebrities . . ."*: Earl Blackwell, President of the Celebrity Information and Research Service, Inc., letter to John Hersey, December 11, 1946, John Hersey Papers, Beinecke Library, Yale University.

141 *"Miss Parsons did . . ."*: Harold Ross letter to John Hersey and William

Shawn, December 16, 1946, *New Yorker* records, New York Public Library.

141 *his . . . "eye-witness account . . ."*: "1946 Pulitzer Prize Winner in reporting: William Leonard Laurence of *The New York Times*," Pulitzer Prizes website: https://www.pulitzer.org/winners/william-leonard -laurence.

141 *Hersey wasn't eligible*: Harold Ross letter to Dorothy Thompson, September 17, 1946, John Hersey Papers, Beinecke Library, Yale University.

141 *one of the notable documents*: Verner W. Clapp letter to John Hersey, September 3, 1946, John Hersey Papers, Beinecke Library, Yale University.

141 *pledge the first draft*: Donald G. Wing letters to John Hersey, November 8, 1946, and January 15, 1947, John Hersey Papers, Beinecke Library, Yale University. *press release announcing the acquisition*: Untitled press release, Yale University News Bureau, May 2, 1947. The top paragraph announced that "John Hersey, Yale 1936, well-known author, has presented the original manuscript for his famous 'Hiroshima' to the Yale University Library."

141 *"How did [Hersey] . . ."*: Harold Ross letter to Harding Mason, May 19, 1947, *New Yorker* records, New York Public Library.

141 *"That book ought . . ."*: Harold Ross letter to Alfred Knopf, September 13, 1946, *New Yorker* records, New York Public Library. *"Widest market possible"*: "Publication Proposal: Hiroshima," September 4, 1946, Alfred A. Knopf, Inc., Archive, Harry Ransom Center, University of Texas at Austin. Knopf estimated that Hiroshima would sell more than 100,000 copies within the first six months.

141 *"Working fast to . . ."*: Harold Ross letter to Dore Schary, September 6, 1946, *New Yorker* records, New York Public Library.

142 *"destined to be . . ."* and *"It is hard . . ."*: Book-of-the-Month advertisement, Fall 1946, *New Yorker* records, New York Public Library.

142 *"because as you . . ."* and *"put you in . . ."*: Randall Gould letter to John Hersey, September 26, 1946, John Hersey Papers, Beinecke Library, Yale University.

CHAPTER SEVEN: Aftermath

143 *"We all exhausted . . ."*: McGeorge Bundy interview with Robert Jay Lifton, 1994, as quoted in Robert Jay Lifton and Greg Mitchell, *Hiroshima in America* (New York: G. P. Putnam's Sons, 1995), 90.

144 *"America forgets so . . ."*: General Thomas Farrell letter to Bernard Baruch, September 3, 1946, as quoted in James Hershberg, *James B. Conant: Harvard to Hiroshima and the Making of the Nuclear Age* (New York: Alfred A. Knopf, 1993), 295.

144 *William Shawn received a letter*: Maj. Cav. Robert J. Coakley to William Shawn, September 23, 1946, *New Yorker* records, New York Public Library.

144 *"aid and control . . ."* and *"the catastrophe of . . ."*: General Leslie Groves, "Remarks of Major General L. R. Groves Before the Command and General Staff School, Fort Leavenworth, Kansas," September 19, 1946, Hoover Institution, Stanford University.

145 *enormous demand . . . within the military*: Maj. Cav. Robert J. Coakley to William Shawn, September 23, 1946, *New Yorker* records, New York Public Library.

145 *"It is my . . ."*: Jay Cassino, Chief, Magazines and Books, Public Relations Division, War Department, letter to R. Hawley Truax, January 8, 1947, *New Yorker* records, New York Public Library.

146 *"Do we know . . ."*: Norman Cousins, "The Literacy of Survival," *Saturday Review of Literature*, September 14, 1946, 14.

146 *"Why reveal a . . ."*: Fleet Adm. William F. (Bull) Halsey Jr. as quoted in "Use of A-Bomb Called Mistake," *Watertown Daily News*, September 9, 1946, 7.

147 *"would sacrifice public . . ."*: Report of the Committee on Political and Social Problems, Manhattan Project "Metallurgical Laboratory," University of Chicago, June 11, 1945, Atomic Heritage Foundation website: https://www.atomicheritage.org/key-documents/franck-report.

147 *"If atomic bombs . . ."*: J. Robert Oppenheimer, farewell speech at Los Alamos, October 16, 1945, as quoted in Kai Bird and Martin J. Sherwin, *American Prometheus: The Triumph and Tragedy of J. Robert Oppenheimer* (New York: Alfred A. Knopf, 2005), 329.

147 *vast potential energy*: Richard Rhodes email to Lesley Blume, January 25, 2020.

147 *disavowed any personal "fatherhood"*: "On the Atomic Bomb, as Told to Raymond Swing, Before 1 October 1, 1945," *Atlantic Monthly*, November 1945, as reprinted in David E. Rowe and Robert Schulmann, eds., *Einstein on Politics: His Private Thoughts and Public Stands on Nationalism, Zionism, War, Peace, and the Bomb* (Princeton, NJ: Princeton University Press, 2007), 376. *horrific destruction*: Albert Einstein letter to Niels Bohr, December 12, 1944, as reprinted in David E. Rowe and Robert Schulmann, eds., *Einstein on Politics*, 364.

147 *"Today, the physicists . . ."*: Albert Einstein, "The War Is Won, but the Peace Is Not," speech, Hotel Astor, New York, December 10, 1945, as reprinted in David E. Rowe and Robert Schulmann, eds., *Einstein on Politics*, 381.

147 *every . . . population on earth vulnerable*: Albert Einstein, "The Real Problem Is in the Hearts of Men," *New York Times Magazine*, June 23, 1946, as reprinted in David E. Rowe and Robert Schulmann, eds., *Einstein on Politics*, 383–84.

148 *"To the village . . ."*: Ibid., 387.

148 *"Mr. Hersey has . . ."* and *"this picture has . . ."*: Albert Einstein letter to recipients of "special facsimile reprint" of "Hiroshima," September 6, 1946, John Hersey Papers, Beinecke Library, Yale University.

148 *returned from . . . holiday* and *alarmed him greatly*: Jennet Conant,

Man of the Hour: James B. Conant, Warrior Scientist (New York: Simon & Schuster, 2017), 377–78, 380.

149 *poison gas* and *"Grand Duke"*: "James B. Conant Is Dead at 84; Harvard President for 20 Years," *New York Times*, February 12, 1978, 36.

149 *"I wept as . . ."* and *"whoopee spirit"*: Arthur Squires to J. Balderston, September 7, 1946, as quoted in Alice Kimball Smith, *A Peril and a Hope: The Scientists' Movement in America: 1945–47* (Chicago: The University of Chicago Press, 1965), 80–81.

149 *"War is ethically . . ."*: James B. Conant letter to Muriel Popper, June 21, 1968, as quoted in Jennet Conant, *Man of the Hour*, 378.

149 *"Monday morning quarterbacking," "all this talk," "It seems to me . . . ,"* and *"group of so-called . . ."*: James B. Conant to Harvey H. Bundy, September 23, 1946, Harvard University, Records of President James Bryant Conant, Harvard University Archives.

149 *recklessly and without . . . consideration*: Harry S. Truman letter to Henry L. Stimson, December 31, 1946, as quoted in James Hershberg, *James B. Conant*, 295.

150 *"The Japanese were . . ."*: Harry S. Truman to Karl T. Compton, December 16, 1946, as reprinted in the *Atlantic*, February 1947.

150 *"I never read . . ."* and *"Just makes me . . ."*: Leonard Lyons item, *New York Post*, October 7, 1946.

150 *to the president's press secretary*: Harold Ross letter to Charles G. Ross, October 9, 1946, *New Yorker* records, New York Public Library.

150 *"The President may . . ."*: Charles G. Ross letter to Harold Ross, October 14, 1946, *New Yorker* records, New York Public Library.

151 *"I have asked . . ."*: Harry S. Truman letter to Karl T. Compton, December 16, 1946, Truman Collection via Martin Sherwin, and as quoted in Michael Yavenditti, "American Reactions to the Use of Atomic Bombs on Japan, 1945–1947," dissertation for doctorate of philosophy, University of California, Berkeley, 1970, 378.

151 *"no one who . . ."*: James B. Conant to Harvey H. Bundy, September

23, 1946, Harvard University, Records of President James Bryant Conant, Harvard University Archives. *memoirs on his Long Island estate*: James Hershberg, *James B. Conant*, 294. *over lunch*: Robert Jay Lifton and Greg Mitchell, *Hiroshima in America*, 98.

151 *"The statement should . . ."*: James B. Conant to Harvey H. Bundy, September 23, 1946, Harvard University, Records of President James Bryant Conant, Harvard University Archives.

151 *"an integrity that . . ."*: "Henry L. Stimson Dies at 83 in His Home on Long Island," *New York Times*, October 21, 1950, 6.

151 *"victim"*: Henry L. Stimson letter to Felix Frankfurter, December 12, 1946, as quoted in James Hershberg, *James B. Conant*, 295. *"awake at night . . ."*: John J. McCloy quoted in Robert Jay Lifton and Greg Mitchell, *Hiroshima in America*, 99.

152 *"I did not . . ."*: Henry L. Stimson, June 6, 1945, as quoted in Monica Braw, *The Atomic Bomb Suppressed* (Armonk, NY: M. E. Sharpe, Inc., 1991), 138.

152 *"the old-boy War Department . . ."*: James Hershberg, *James B. Conant*, 294.

152 *"a splendid description . . ."*: General Leslie Groves, "Comments on Article by Secretary of War, Henry L. Stimson, *Harper's* Magazine, February 1947, Explains Why We Used the Atomic Bomb," undated, National Archives and Records Administration. In these comments, General Groves reiterates that he had gone over and inputted on the Stimson article when it was in a near-final draft.

152 *Such an article was necessary*: General Leslie Groves letter to McGeorge Bundy, November 6, 1946, as quoted in Robert Jay Lifton and Greg Mitchell, *Hiroshima in America* (New York: G. P. Putnam's Sons, 1995), 99.

152 *"staggering report"*: John Chamberlain, "The New Books," *Harper's*, December 1, 1946.

153 *"mere recital of . . ."*: James B. Conant, as quoted in James Hershberg, *James B. Conant*, 296.

152 *"eliminate all sections . . ."*: James B. Conant letter to McGeorge Bundy, November 30, 1946, as quoted in Jennet Conant, *Man of the Hour*, 388.

152 *"the propaganda against . . ."*: James B. Conant letter to Henry L. Stimson, as quoted in Jennet Conant, *Man of the Hour*, 389.

152 *"I have rarely . . ."*: Henry L. Stimson letter to Felix Frankfurter, December 12, 1946, as quoted in James Hershberg, *James B. Conant*, 295.

152 *"I think you . . ."* and *"straighten[ing] out the record . . ."*: Harry S. Truman letter to Henry L. Stimson, December 31, 1946, as quoted in Robert Jay Lifton and Greg Mitchell, *Hiroshima in America*, 102.

152 *"rather difficult class . . ."*: Stimson as quoted in Robert Jay Lifton and Greg Mitchell, *Hiroshima in America*, 102, and Michael J. Yavenditti, "John Hersey and the American Conscience," *Pacific Historical Review* 43, no. 1 (February 1974), 44.

152 *sent a copy . . . to Henry Luce*: Robert Jay Lifton and Greg Mitchell, *Hiroshima in America*, 102.

154–156 *"our least abhorrent . . . ,"* *"save[d] many times . . . ,"* *"strangling blockade,"* *"In recent months . . . ,"* *"all who may . . . ,"* *"to a race . . . ,"* *"might be expected . . . ,"* *"discarded as impractical,"* *"nothing would have . . . ,"* *"dread of many . . . ,"* *"a weapon of . . . ,"* *"psychological weapon,"* and *"All of the . . ."* : Henry L. Stimson, "The Decision to Use the Atomic Bomb," *Harper's*, February 1947.

155 *40,000 American military deaths and 150,000 wounded*: James Hershberg, *James B. Conant*, 301.

156 *"it cannot be . . ."*: "The United States Strategic Bombing Survey: The Effects of Atomic Bombs on Hiroshima and Nagasaki," June 30, 1946.

156 *"a revolutionary character,"* *"generally unfamiliar nature,"* *"as legitimate as . . . ,"* and *"no man in . . ."*: Henry L. Stimson, "The Decision to Use the Atomic Bomb."

157 *"very well"*: Harry S. Truman letter to Henry L. Stimson, as quoted in Jennet Conant, *Man of the Hour*, 389.

157 *"We deserve some . . .":* McGeorge Bundy letter to Henry L. Stimson, as quoted in Jennet Conant, *Man of the Hour,* 391.

157 *declined to give significant coverage:* "National Affairs: 'Least Abhorrent Choice,'" *Time,* February 3, 1947, 20.

158 *"the German military . . ."* and *"that the most . . .":* "War and the Bomb," *New York Times,* January 28, 1947, 22.

158 *"brought death to . . . ," "gloss over it,"* and *"The face of . . .":* Henry L. Stimson, "The Decision to Use the Atomic Bomb."

159 *new printing of 1 million:* Louis Forster Jr. letter to John Hersey, February 14, 1947, *New Yorker* records, New York Public Library.

159 *"Besides [the Knopf] . . .":* "Hersey's 'Hiroshima' One Year Afterward," Knopf press release, November 20, 1947, John Hersey Papers, Beinecke Library, Yale University.

160 *"Christ knows what . . .":* Harold Ross letter to John Hersey and William Shawn, November 25, 1947, *New Yorker* records, New York Public Library.

160 *"imperialist," "expansionist plans," "shed their invaluable,"* and *"monopolistic possession of . . .":* Speech delivered by V. M. Molotov, Head of the Soviet Delegation to the United Nations Organization, Minister of Foreign Affairs of the U.S.S.R., on "The Soviet Union and International Cooperation," General Assembly of the United Nations, New York City, October 29, 1946.

161 *"to show who . . . ," "were not aimed . . . ,"* and *"They said, bear . . .":* V. M. Molotov, *Molotov Remembers: Inside Kremlin Politics,* edited by Albert Resis (Chicago: Ivan R. Dee, Inc., 1993), 55, 58.

161 *"attempt to intimidate . . .":* "Atomic Race," *New York Herald Tribune,* August 30, 1946, 14.

161 *On August 23, Mikhail Ivanav, consul:* David Holloway, *Stalin and the Bomb: The Soviet Union and Atomic Energy, 1939–1956* (New Haven, CT: Yale University Press, 1994), 129–30.

161 *"the bomb constituted . . .":* Alexander Werth, *Russia at War, 1941–1945,* 925, as quoted in David Holloway, *Stalin and the Bomb,* 127.

162   *"relatively at ease . . ."*: John Hersey, handwritten notation on draft of Raoul Fleischmann letter to Andrei Gromyko, undated (but between December 6, 1946, when first dated draft is on file, and December 12, 1946, when the letter was finalized), *New Yorker* records, New York Public Library.

162   *"not a word . . ."*: John Hersey, "Engineers of the Soul," *Time*, October 9, 1944, as quoted in Nancy L. Huse, *The Survival Tales of John Hersey* (Troy, NY: The Whitson Publishing Company, 1983), 32.

162   *"Mr. Nyet," "Grim Grom,"* and *"Old Stoneface"*: David Remnick, "Gromyko: The Man Behind the Mask," *Washington Post*, January 7, 1985.

162–63   *"American version of . . ."*: Oskar Kurganov, *Amerikantsy v Iaponii* (*Americans in Japan*), (Moscow: Sovetskii Pisatel', 1947), 11. *"little fright"* and *"no such 'atomic'. . ."*: Ibid., 53–54. Translated from Russian by Anastasiya Osipova.

163   *"Russian reply to . . ."*: Joseph Newman, "Soviet Writer Scoffs at Power of Atom Bomb; Says Nagasaki Destruction Was Not Nearly So Bad as Was Claimed by U.S." *New York Herald Tribune*, July 14, 1947.

163   *"relishes the torment . . ."* and *"spread panic"*: A. Leites, "O zakonakh istorii i o reaktsionnoj isterii" ("On the Laws of History and Reactionary Hysteria"), *Pravda*, 178th ed., July 12, 1947, 2–3. Translated from Russian by Anastasiya Osipova.

163   *"military spirit"* and *"propaganda of aggression . . ."*: R. Samarin, "Miles Americanus," *Soviet Literature*, 1949, as translated and excerpted in "A Cold War Salvo," Harry Schwartz, *New York Times Book Review*, July 17, 1949, 81.

164   *The official reason given*: Office Memorandum, United States Government, from A. H. Belmont to C. H. Stanley, June 2, 1950, the Records of the Federal Bureau of Investigation. The author of this memorandum noted that Arthur Hersey, "brother of John Richard Hersey, was the subject of a loyalty investigation conducted in 1948 . . . conducted by the Bureau . . . based on information that Arthur B. Hersey was

treasurer of the Washington Committee for Aid to China in 1941, which organization was cited by the House Committee on Un-American Activities as a Communist Front." Regarding Hoover directing interview of John Hersey: Internal teletype, J. Edgar Hoover, June 2, 1950, the Records of the Federal Bureau of Investigation. Hoover cites Hersey's FBI file number as 121-668 and asks that a subsequent report on Arthur Hersey be supplemented with "appropriate information [from] your files, re: John Hersey."

164 *"obviously and quite . . .":* "Results of Investigation: Loyalty of Government Employees: Arthur Baird Hersey," June 19, 1950, 1, the Records of the Federal Bureau of Investigation. *$10 contribution to the ACLU:* Federal Bureau of Investigation, "Report title: Arthur Baird Hersey, Economist, Board of Governors of the Federal Reserve System, Washington, D.C.," File no. 121-70, June 14, 1950, 4, the Records of the Federal Bureau of Investigation.

164 *"strong and lasting":* Hersey lecture quoted in "Results of Investigation: Security of Government Employees: Arthur Baird Hersey," May 25, 1954, 1, the Records of the Federal Bureau of Investigation.

164 *FBI officials came to interview him:* "Results of Investigation: Loyalty of Government Employees: Arthur Baird Hersey," June 19, 1950, 3, the Records of the Federal Bureau of Investigation. Hersey interviewed about his 1946 reporting trip to Japan: Federal Bureau of Investigation, "Report title: Arthur Baird Hersey, Economist, Board of Governors of the Federal Reserve System, Washington, D.C.," File no. 121-70, June 14, 1950, 2, the Records of the Federal Bureau of Investigation.

165 *informed . . . that SCAP was blocking . . . the book:* Carl Mydans to John Hersey, as recounted in John Hersey letter to William Koshland, December 16, 1946, Alfred A. Knopf, Inc., Archive, Harry Ransom Center, University of Texas at Austin.

165 *told by a reverend:* Reverend Calvert Alexander letter to Charlotte Chapman of the *New Yorker*, September 16, 1946, *New Yorker* records, New York Public Library.

166   *"I didn't know . . . ,"* *"everything in [it] . . . ,"* and *"remember all the . . ."* : Dr. Masakazu Fujii and Hatsuyo Nakamura, as quoted in Norman Cousins, "John Hersey," *Book-of-the-Month Club News,* March 1950.

166   *"best thanks for . . ."* and *"Suppose how I . . ."*: Dr. Masakazu Fujii postcard to John Hersey, *New Yorker* records, New York Public Library.

166–67 *"greatly surprised and . . . ,"* *"It is a . . . ,"* and *"interviewed"*: Reverend Kiyoshi Tanimoto letter to John Hersey, March 8, 1947, *New Yorker* records, New York Public Library.

167   *"without the slightest . . ."* and *"They stem from . . ."*: General Douglas MacArthur cable to Oscar Hammerstein, April 6, 1948, MacArthur Memorial Archives, RG-5, Box 6, Fol. 3, and also released as a General Headquarters press release on April 7, 1948, MacArthur Memorial Archives, RG-25: Addresses, Speeches, Box 1.

167   *an immediate bestseller*: Ralph Chapman, "All Japan Puts 'Hiroshima' on Best Seller List," *New York Herald Tribune,* May 29, 1949. Chapman reported that "A first edition of 40,000 copies was sold out within two weeks of publication and a second printing of 10,000 is now being prepared."

167   *"expresses a humanism . . ."*: Review of Hersey's Japanese-language *Hiroshima, Tokyo Shimbun,* May 8, 1949.

167   *"Hold a reunion . . ."*: Harold Ross letter to John Hersey, December 9, 1946, *New Yorker* records, New York Public Library.

167   *essentially gone underground*: The *New Yorker* did, however, in June 1947, publish a work of short fiction of Hersey's: "A Short Wait," in the June 14, 1947, issue.

168   *"probably about forty-two . . ."*: Harold Ross letter to Jack Wheeler, May 22, 1950, *New Yorker* records, New York Public Library.

168   *"outrage at human . . ."* and *"the experience gave . . ."*: John Hersey, "The Mechanics of a Novel," *Yale University Gazette,* July 1952.

169   *"would be to . . ."*: William Shawn letter to Harold Ross, January 11, 1949, *New Yorker* records, New York Public Library.

Notes

169 *666 times as powerful*: Matt Korda email to Lesley Blume, December 5, 2019.

169 *as elaborate a statement*: "Harry Truman: 'The Japanese Were Given Fair Warning,'" *Atlantic*, February 1947.

170 *"There's nothing new . . ."*: John Hersey, "Profiles: Mr. President: I. Quite a Head of Steam," *New Yorker*, April 14, 1951, 49.

170 *"regarded as the . . ."*: John Hersey, "Profiles: Mr. President: V. Weighing of Words," *New Yorker*, May 5, 1951, 36.

170 *"I don't think . . ."* and *"That's just what . . ."*: John Hersey, "Profiles: Mr. President: V. Weighing of Words," *New Yorker*, May 5, 1951, 37.

170 *"the government had . . ."*: John Hersey, "The Wayward Press: Conference in Room 474," *New Yorker*, December 16, 1950, 86.

171 *"To my mind . . ."* and *"I don't care . . ."*: John Hersey, "Profiles: Mr. President: I. Quite a Head of Steam," *New Yorker*, April 14, 1951, 46.

171 *"active consideration"*: John Hersey, "The Wayward Press: Conference in Room 474," *New Yorker*, December 16, 1950, 85.

171 *valuable cost-saving device*: "For Eisenhower, 2 Goals If Bomb Was to Be Used," *New York Times*, June 8, 1984, A8.

172 *"bombings of Hiroshima . . ."*: Martin Sherwin interview with Lesley Blume, March 22, 2018. Sherwin had been dismayed by Hersey's statement: "My view is exactly the opposite, that [the Hiroshima and Nagasaki bombings] started the arms race. I was not thrilled with his view."

172 *"I think that . . . ,"* *"The memory of . . . ,"* *"what it would . . . ,"* *"very spotty in . . . ,"* *"must never have . . . ,"* and *"The control of . . ."*: John Hersey, "John Hersey: The Art of Fiction No. 92," interview by Jonathan Dee, *Paris Review*, issue 100, Summer–Fall 1986

173 *"astonishing resources for . . ."* and *"In spite of . . ."*: John Hersey, *Here to Stay*, vii.

Notes

EPILOGUE

175 *"gentle despot"*: Eric Pace, "William Shawn, 85, Is Dead; *New Yorker*'s Gentle Despot," *New York Times*, December 9, 1992, 1.

175 *"It was quite . . ."*: John Hersey, "John Hersey, The Art of Fiction No. 92," interview by Jonathan Dee, *Paris Review*, no. 100, Summer–Fall 1986.

176 *"He was a . . ."*: Baird Hersey as quoted in Russell Shorto, "John Hersey, The Writer Who Let 'Hiroshima' Speak for Itself," *New Yorker*, August 31, 2016.

176 *"one worried grandpa"* and *"figure of the . . ."*: John Hersey, "The Art of Fiction No. 92."

176 *"gaudy phoenix had . . ."* and *"city of strivers . . ."*: John Hersey, "Hiroshima: The Aftermath," *New Yorker*, July 15, 1985, 47.

176 *gave 582 lectures*: Associated Press, "Kiyoshi Tanimoto Dies; Led Hiroshima Victims," reprinted in *New York Times*, September 29, 1986, B14.

176 Quotes from Reverend Tanimoto's Senate prayer: *Congressional Record: Proceedings and Debates of the 82nd Congress*, vol. 97, part 16.

177 *"the high point . . ."*: John Hersey, "Hiroshima: The Aftermath," 59.

177 *it turned out*: When introducing Reverend Tanimoto to his audience, show presenter Ralph Edwards stated on the air that he had actually worked "for weeks" with several people to coordinate the Tanimoto show, including Hersey—something that Hersey did not mention in his reporting on Reverend Tanimoto's *This Is Your Life* appearance in his "Hiroshima: The Aftermath" reporting for the *New Yorker*.

177 *"You thought, of course . . ."* and *"retell the story . . ."*: Ralph Edwards, *This Is Your Life*," NBC, May 11, 1955, https://www.youtube.com/watch?v=KPFXa2vTErc.

178 *reached out for his hand*: Koko Tanimoto Kondo interview with Lesley Blume, November 29, 2018.

178 *"he had gone . . ."*: John Hersey, "Hiroshima: The Aftermath," 61.

178 *$50,000 in donations*: John Hersey, "Hiroshima: The Aftermath," 62.

178 *died in 1986*: "Kiyoshi Tanimoto Dies; Led Hiroshima Victims," *New York Times*, September 29, 1986, B14.

178 *"When Hiroshima was . . ."*: Father Wilhelm Kleinsorge and Matoko Takakura, "The First Interviews with Atomic Bomb Victims," *Asahigraph*, August 6, 1952, translated from Japanese by Ariel Acosta.

178 *"A-bomb cataract," "a living corpse,"* and *"the only two . . ."*: John Hersey, "Hiroshima: The Aftermath," 49, 50.

178 *"After being written . . ."* and *"I struggled emotionally . . ."*: Dr. Masakazu Fujii, "The First Interviews with Atomic Bomb Victims," *Asahigraph*.

179 "HERE IS FUJII . . .": Ray C. Anderson, Ph.D., M.D., *A Sojourn in the Land of the Rising Sun: Japan, the Japanese, and the Atomic Bomb Casualty Commission: My Diary, 1947–1949* (Sun City: Elan Press, 2005), 428.

179 "DR. FUJII HERE . . .": Leonard Gardner letter to John Hersey, December 30, 1951, John Hersey Papers, Beinecke Library, Yale University.

179 *"It's become one . . ." :* Dr. Masakazu Fujii, as quoted in Norman Cousins, "John Hersey: Journalist into Novelist," *Book-of-the-Month Club News*, March 1950.

179 *"a cancer the size of a Ping-Pong ball . . ."*: John Hersey, "Hiroshima: The Aftermath," 56.

179–80 *"There was a . . ."* and *"I don't want . . ."*: Dr. Terufumi Sasaki, "The First Interviews with Atomic Bomb Victims," *Asahigraph*.

180 *"For four decades . . ."* and *"[H]is one bitter . . ."*: John Hersey, "Hiroshima: The Aftermath," 42, 47.

180 *"I wouldn't say . . ."*: Sasaki Toshiko, "First Interviews with Atomic Bomb Victims," *Asahigraph*, August 6, 1952, translated from Japanese by Ariel Acosta.

180 *"a pattern of . . ."*: John Hersey, "Hiroshima: The Aftermath," *New Yorker*, 53.

180 *"It is as if I had been given . . ."*: Sister Dominique Sasaki, as quoted in John Hersey, "Hiroshima: The Aftermath," *New Yorker*, 54.

181 *"so afraid of . . ."*: Hatsuyo Nakamura, "The First Interviews with Atomic Bomb Victims," *Asahigraph.*

181 *William Shawn was ousted* and *died of a heart attack*: Eric Pace, "William Shawn, 85, Is Dead; New Yorker's Gentle Despot," *New York Times*, December 9, 1992, 1, 39.

181 *Hersey died of cancer*: Hendrik Hertzberg, "John Hersey," *New Yorker*, April 5, 1993, 111, and Richard Severo, "John Hersey, Author of 'Hiroshima,' Is Dead at 78," *New York Times*, March 25, 1993, A1 and B11.

181 *nearly 3 million inhabitants*: Hiroshima official tourism website, http://visithiroshima.net/about/.

181 *majority of those surveyed*: A UCLA study analyzing reader feedback to "Hiroshima" calibrated that "53% of the writers believe the story is a contribution to public good." Source: "Reaction to John Hersey's 'Hiroshima' Story," Joseph Luft, UCLA report, July 14, 1947, *New Yorker* Papers, The New York Public Library.

182 *"fighting to be . . ."*: George Weller: *First into Nagasaki: The Censored Eyewitness Dispatches on Post-Atomic Japan and Its Prisoners of War*, Anthony Weller, ed. (New York: Three Rivers Press, 2006), 274.

182 *"the very things . . ."*: "John Hersey, The Art of Fiction No. 92."

183 *"escape into easy . . ."*: Albert Einstein, "The War is Won, But the Peace is Not," speech, December 10, 1945, as reprinted in *Einstein on Politics: His Private Thoughts and Public Stands on Nationalism, Zionism, War, Peace, and the Bomb*, ed. David E. Rowe and Robert Schulmann (Princeton, NJ: Princeton University Press, 2007), 382.

Index

Pages beginning with 193 refer to notes.

Index

Groves, Leslie R. *(continued)*
  on Soviet acquisition of atomic
    bomb, 169
  Stimson article and, 152

Halsey, William F. "Bull," Jr.:
  atomic bombing of Japan seen as
    mistake by, 146
  on Japanese peace feelers, 155
Harper's, 152, 157, 159
  Stimson article cover of, 153–54
Harrison, George L., 152
Hersey, Arthur, 251–52
Hersey, Baird, 176
Hersey, Frances Ann, 17, 37
Hersey, John, 1–3, 5, 8–10, 141
  anti-Japanese prejudices of, 53–54
  cables to Shawn from, 54–55, 59, 69
  in call for U.S.-Soviet friendship,
    164
  as celebrity, 140
  death of, 181
  in decision to focus on human
    effects of bomb, 47–48, 58, 82,
    95, 122
  early career of, 16–17
  education of, 17–18
  FBI investigation of, 164–65,
    251–52
  as fearing for humanity's future, 14,
    21–22, 72
  flu contracted in Manchuria by,
    57–58
  "Hiroshima" follow-up article
    initially rejected by, 167–68
  on humans' will to survive, 173
  on legacy of "Hiroshima," 171–73
  Life and, 53, 60
  Luce's relationship with, 18–19, 36,
    41, 136
  military-friendly stories by, 42
  moral compass of, 17

Nagasaki bombing viewed as
  criminal act by, 22
1946 China reporting by, 52, 54–55,
  57
optimism of, 168, 173
planned Asia trip of, 42
in planning of Hiroshima trip,
  49–52
possible Hiroshima story discussed
  by Shawn and, 42–43
publicity shunned by, 17, 127, 176,
  198–99
as Pulitzer Prize winner, 16–17, 61,
  140
reprint income from "Hiroshima"
  donated to Red Cross by, 123
in retreat to North Carolina, 127,
  135
in return to New York from Japan, 97
SCAP's approval given for Japan
  trip of, 59
on Shawn's editing genius, 41,
  112–13
Shawn's warning to, about not filing
  story from Japan, 65
in shift from reporting to fiction-
  writing, 168
as *Time* reporter in Moscow, 18, 51,
  66, 161, 162, 164, 168
as war correspondent, 15–16, 18,
  37, 48, 51, 75, 168
as war hero, 66
Hersey, John, in Hiroshima, 9–10, 71–96
  blast survivors interviewed by,
    87; *see also* "Hiroshima"
    protagonists
  Fujii interviewed by, 91–93
  impact of devastated landscape on, 72
  Kleinsorge as translator for, 80, 90
  Kleinsorge interviewed by, 77–80
  Mrs. Nakamura interviewed by,
    87–89

Index

Hiroshima, atomic bombing of
(continued)
  Terufumi Sasaki's eyewitness
    account of, 90–91
  Toshiko Sasaki's eyewitness account
    of, 94
  Toshio Nakamura's recollection of,
    106–7
  U.S. government cover-up of, 28–33
  U.S. government documentation of,
    62–64
  U.S. government studies of, 119
  as war crime, 129
  windstorms in wake of, 79, 83–84, 89
Hiroshima (Hersey; book), 11, 141–42
  as banned in Japan, 165
  censorship of, 160–63
  international editions of, 159–60
  MacArthur's approval of Japanese
    distribution of, 167
  photographs of protagonists on back
    cover of, 159
  possibility of Russian edition of,
    161–62
Hiroshima (radio adaptation), 136–38
"Hiroshima" (Hersey; *New Yorker*
    article), 1, 2
  August 29, 1946 publication of, 125
  as contribution to public good, 134,
    181
  detailed description of radiation
    poisoning in, 103–4, 121
  as deterrent to nuclear war, 5–7,
    11–13
  as document of conscience, 121
  editor's page-one note to, 126, 211
  effects of bombing on humans as
    focus of, 47–48, 99, 122, 159
  famous lede of, 100
  government cover-up of bomb
    effects revealed by, 127, 145–46,
    148

Hersey's reprint income from
    donated to Red Cross, 123
  intense editing process for, 109–14
  intentional suppression of horror
    in, 99
  interweaving of protagonists' stories
    in, 99
  legacy of, 5, 140–42, 157–60,
    171–73, 181–82
  mock-up of magazine issue, 122–24
  as needing to be flawless, 114
  negative responses to, 134–36
  newsstand sales of, 125, 132
  novelistic style of, 98–99
  100,000 death toll figure in, 7,
    113–14, 155
  as possible PR for U.S. nuclear
    superiority, 119–20
  post-War Department meeting
    changes to, 120, 236
  pre-publication secrecy surrounding,
    10, 110–11, 232–33
  press response to, 10, 127–32,
    134–36, 146
  public opinion and, 148
  published in book form, *see*
    Hiroshima (Hersey; book)
  reader responses to, 133–36
  reader's loss of sense of time in, 111
  reprints of, 123, 128, 145, 148, 165
  "restricted data" standard and,
    115–18
  as "scoop of the century," 10
  seen as antipatriotic propaganda,
    134
  seen as U.S. propaganda by Soviets,
    163
  single-issue publication of, 108
  Soviet response to, 160–63
  stripped-down language of, 99
  true nature of nuclear weapons
    revealed by, 128

Index

Index

Index

Shanghai, China, 52, 54–55, 57, 59
Shanghai Evening Post (China),
    142
Shawn, William, 135
    death of, 181
    in decision to focus on human
        effects of bomb, 47–48, 58,
        122
    in decision to run "Hiroshima" in
        single issue, 107–9
    in decision to submit "Hiroshima"
        for War Department approval,
        117–19
    early copies of "Hiroshima" sent to
        major newspapers by, 126–27
    editing style of, 98
    and enormous demand for
        "Hiroshima" copies in military,
        144
    Hersey on editing genius of, 41,
        112–13
    Hersey's cables to, 54–55, 59, 69
    "Hiroshima" proofs delivered to
        printer by, 123–24
    as *New Yorker* deputy editor, 37–41
    as *New Yorker* editor, 175
    ousted as *New Yorker* editor, 181
    possible Hiroshima story discussed
        by Hersey and, 42–43
    and press response to "Hiroshima,"
        132
    retiring personality of, 38–39, 111
    in warning to Hersey not to file
        story in Japan, 65
    and wartime refocusing of *New
        Yorker*, 10, 231
Shigeto, Fumio, 86
Siemes, Johannes:
    at Asano Park, 86
    eyewitness account of Hiroshima
        bombing by, 64, 75, 78–79, 80,
        83, 217, 222–23

Society of Jesus, Central Mission and
    Parish House of, 75–77, 79–80,
    88–89, 92
    see also Jesuit priests
Solomon Islands, 16, 37
Soviet Union:
    attempted debunking of Hiroshima
        article by, 11
    and Japanese surrender attempts, 155
    as lacking institutional memory, 172
    nuclear weapons acquired by, 12, 169
    possibility of *Hiroshima* edition in,
        161–62
    press accounts of atomic bombing
        suppressed in, 161
    U.S. relations with, 18, 43, 55
    see also Cold War
Stalin, Joseph, 161
statistics, numbing effect of, 6–7
Stimmen der Zeit (Voices of the
    Times), 78
Stimson, Henry L., 149
    "Decision to Use the Atomic Bomb"
        written by, 151–56
    as Truman's senior advisor on
        nuclear energy uses, 151
Strategic Bombing Survey, U.S., 63,
    105–6, 119–21
    on Japanese decision to surrender,
        156
Sussan, Herbert, 62
    Hersey's meeting with, 63–64, 75
Swing, Raymond, 138

Takakura, Makoto, *see* Kleinsorge,
    Wilhelm
Tanimoto, Chisa, 81–82, 83, 84–85,
    177–78
Tanimoto, Kiyoshi, 100, 101
    as antinuclear advocate, 176–77
    in attempted rebuilding of church,
        80–81

Index

Index

United States:
anticommunist fervor in, 164–65
United States *(continued)*
and future use of nuclear weapons,
146
homecoming of troops in, 43
It Can't Happen Here and, 182
Japanese internment camps in, 8
North Korea and, 13–14
nuclear monopoly of, 56, 103, 105,
119, 160
nuclear policy of, 12–13
Soviet relations with, 18, 43, 55
Soviet Union and, *see* Cold War
Tanimoto in, 80, 176–78
uranium bomb, *see* atomic bomb

Victory in Europe Day (V-E Day), 15,
17, 19
Victory over Japan Day (V-J Day),
23–24

Wall, The (Hersey), 168
war correspondents, 20, 26, 39–41,
Hersey as, 15–16, 18, 37, 51, 75,
168
see also press, western, in Japan
war crimes, 43, 48, 68, 96, 106
atomic bombing of Japan as, 129
War Department, U.S., 19, 22, 51,
55–56, 131, 144
censorship by, 10, 40–41, 115–18, 120
"Hiroshima" publication approved
by, 121, 236–37

pre-publication submission
"Hiroshima" to, 117–21
Warsaw Ghetto, 168
Weinberger, Caspar, 172
Weller, George, 51, 60, 67, 103, 182
"lost" Nagasaki reporting by, 32, 49
West, Rebecca, 109
White, E. B., 3, 108
Wilder, Thornton, influence on
"Hiroshima" of, 58–59, 87
Winchell, Walter, 17
Winnacker, Rudolph A., 152
Wolfe, Tom, 176
World War I, 148
World War II, 75
aftermath of, 5, 19, 29, 42
casualties of, 6–7
German surrender in, 15, 17
homecoming of U.S. troops
from, 43
Japanese surrender in, 3, 10, 23, 49,
151
Pacific theater in, 16, 19–20
Potsdam Conference in, 21
press coverage of, 39–41, 182;
see also war correspondents
see also Allied Powers; Axis
Powers

Yale University, 17, 141, 164, 199
Yokohama, Japan, 31, 60, 62
Yoshiki, Satsue, 178
Yuzaki, Hidehiko, 2
Yuzaki, Minoru, 194